A reconciliation of theories of the very small and the very large scale is one of the most important single issues in physics today. Many people today are unaware that back in the 1930s, Sir Arthur Eddington, the celebrated astrophysicist, made great strides towards his own 'theory of everything'.

In 1936 and 1946 Eddington's last two books were published. Unlike his earlier lucid and authoritative works, these are strangely tentative and obscure – as if he were nervous of the significant advances he might be making. This volume examines how Eddington came to write these uncharacteristic books – in terms of the physics and history of the day – and what value they have to modern physics. The result is an illuminating description of the development of theoretical physics in the first half of the twentieth century from a unique point of view: how it affected Eddington's thought. This will provide fascinating reading for scholars in the philosophy of science, theoretical physics, applied mathematics and the history of science.

Eddington's search for a fundamental theory

A key to the universe

Eddington's search for a fundamental theory

A key to the universe

C. W. Kilmister

Formerly Professor of Mathematics, King's College London

CAMBRIDGE
UNIVERSITY PRESS

CAMBRIDGE UNIVERSITY PRESS
Cambridge, New York, Melbourne, Madrid, Cape Town, Singapore, São Paulo

Cambridge University Press
The Edinburgh Building, Cambridge CB2 2RU, UK

Published in the United States of America by Cambridge University Press, New York

www.cambridge.org
Information on this title: www.cambridge.org/9780521371650

First published 1994
This digitally printed first paperback version 2005

A catalogue record for this publication is available from the British Library

ISBN-13 978-0-521-37165-0 hardback
ISBN-10 0-521-37165-1 hardback

ISBN-13 978-0-521-01728-2 paperback
ISBN-10 0-521-01728-9 paperback

Contents

Preface   ix

1 *The mystery*   1

PART 1   1882–1928   9
2 *The astrophysicist*   11
3 *General relativity*   26
4 *Consequences of general relativity*   42
5 *'Something has slipped through the net'*   65
6 *Quantum mechanics*   79

PART 2   1928–33   99
7 *Algebra to the fore: **136***   101
8 *Electric charge: **137***   124
9 *The proton–electron mass-ratio*   156

PART 3   1933–44   185
10 *The turning point*   187
11 *Critical views of RTPE*   207
12 *The last decade*   223

References   250
Index   254

Preface

Most physicists have no difficulty in seeing physics as a single subject. Yet this view, which was straightforwardly tenable until the end of the nineteenth century, is radically inconsistent with the situation since then. There has been a divorce between the theories of the very small and the large scale. Amongst those worried about this the response has been to search for a 'theory of everything'. This phrase has many closely related connotations and to determine which, if any, is the correct one is an inspiring and useful task, not least for the unexpected by-products. So far, however, it has proved a task without any successful outcome. In this book I draw attention to an alternative. Unnoticed by many today, Eddington in the 1930s made great strides towards a different solution of the enigma.

It is half a century since I first succumbed to the Eddingtonian magic – I paraphrase Thomas Mann's phrase to try to do justice to my youthful if uncritical absorption in *Relativity Theory of Protons and Electrons*, which Eddington had published five years or so earlier, in 1936. I had already enjoyed his authoritative *Mathematical Theory of Relativity* with no more difficulty than that produced by the complex mathematical techniques which were new to me. Looking back on it, it surprises me that I could take in without a qualm so many of the unorthodox philosophical views in that book. But *Relativity Theory of Protons and Electrons* was a different matter. Another clutch of mathematical techniques was not enough to obscure a radically new position.

It was very much a book of the 1930s. In the first part of that decade, when the book was gradually coming together, the situation in theoretical physics had become very puzzling. On the one hand, the discoveries of special and general relativity in 1905 and 1915 had been wholly absorbed and largely understood. But understood only within the macroscopic bounds set by the theories themselves. They were confidently expected to explain cosmology in due course, they were known to describe more local astronomical phenomena more closely than Newtonian mechanics and for ordinary mechanics they reduced to Newton's system. On the other hand, the microscopic world had undergone two revolutions. One was at the turn of the century when Planck and Einstein initiated the so-called old quantum theory, a tool-bag of rules for calculating the frequencies of sharp lines in atomic spectra. The second was in 1925–6 when the manifold confusions into which the old quantum theory had fallen were, it seemed, removed not by one but two initiatives, an algebraic one (Heisenberg, Born, Jordan and Dirac) and an analytic one (Schrödinger). The fact that the two were then proved to be substantially equivalent and so were both called *the* new quantum theory was held to confirm their rightness. The puzzling character of physics was that there seemed to be no relation at all between the macroscopic and the microscopic theories, even to the extent that the new quantum theory was consistent with Newtonian mechanics instead of with special relativity (let alone general relativity).

At first Eddington was content to see the two sides as simply two alternative ways of looking at the world, wholly independent of each other. He played no active role in developing the new quantum theory. But in 1928 Dirac's publication of his equation for the electron, an equation which was a natural development of the new quantum theory and yet was consistent with special relativity, alerted Eddington to the problem, a problem that he soon came to see as an opportunity. The realisation came in a personally painful way, for the equation contradicted a folk-belief, strongly held by Eddington amongst the majority, that relativity had in its possession a mathematical device (the tensor calculus) which could churn out all possible equations consistent with its tenets. Dirac's equation was not of this form.

Eddington set to work at once in a way that was characteristic; he employed a number of illuminating models to guide his mathematics. The principal such model was Maxwell's electrodynamics and the way in which Maxwell, by combining the hitherto related but separate theories of magnetism and electricity, was led to the prediction of the value of a physical constant (the speed of light in vacuum) from the known scale constants that related electric and magnetic units. This model was then interpreted in an enormously generalised way. The theories to be linked became relativity and quantum theory, separate indeed but scarcely related at all. The much harder task of joining them would, if carried out successfully, bring even more in the way of prizes. As the work went on the prizes took the form of four physical constants, but now of the dimensionless kind, the most striking ones being the so-called fine–structure constant $e^2/\hbar c$, for which Eddington finally settled on the value 1/137 and the proton–electron mass ratio m_p/m_e for which he gave 1848. At first Eddington's contemporaries were interested but then scepticism took over. In the absence of any other plausible unification of relativity and quantum theory, for quantum electrodynamics was still far in the future, the problem came to be simply ignored, and Eddington's attempted solution with it. Meanwhile Eddington had progressed to a new position. The two theories were not to be seen as parts of a single whole, as in the Maxwell model, but as alternative descriptions which could, just occasionally, apply to the same situation and the results compared. Out of such a comparison could come numerical values of physical constants.

In 1941 much of this background was unknown to me and I was just fascinated by the book on its own terms. As time went on I learnt more and more of the difficulties and the task became that of reconstructing Eddington's work in such a way as to avoid them. In 1945 *Fundamental Theory* appeared after Eddington's death. It was a disappointment to me, for it did not seem to address the real obscurities of the earlier book. I spent a good deal of time clearing up the algebraic aspects of the theory but when this had been done the basic ideas were not much clearer. Yet they continued to dazzle: flashes of insight grouped round a frame of numerical results. I kept returning to Eddington's work and puzzling over it between the other enterprises that filled my working life. I owe a debt of great gratitude to Ted Bastin who often gave me guidance and help over Eddington during this forty-five years.

Eventually I realised that the time was passed in which I personally could hope to unravel the whole enigma. Yet the problem tackled by Eddington, if without complete success, the seemingly unbridgeable gap between quantum mechanics and relativity, particularly the general theory, remained, and I write that in full knowledge of the recent attempts at 'quantum gravity' of all kinds. A problem still pursued me but now it was a different one. How could it have come about that Eddington, such a lucid writer both professionally and in popular science, wrote two such obscure books at the end of his life? This book is an attempt to get behind

Eddington's printed page to answer that question. I found that it needed answers on many levels. The history of science, general history, Eddington's personal circumstances and his peculiar philosophy of science all had their part to play. I had to range over much of twentieth century physics and over Eddington's other work. There was virtually no documentary evidence to go on beyond his printed papers and books, so the reconstruction of his development has been an imaginative journey and the book an intellectual biography. At the end of it I was excited to find a clear picture emerge of how the two books came about.

I have tried throughout to keep technical details to a minimum and to explain most of the physics from first principles, so that the story is as accessible as possible. Yet this is not only a popular book; it is also meant to stimulate others to the task that I have laid down and, if I am fortunate, to give them the benefit of my experience.

Lewes, East Sussex   C. W. K.

1

The mystery

This book is an attempt to unravel a mystery about the writing of two scientific books by Sir Arthur Eddington, his *Relativity Theory of Protons and Electrons* of 1936 and his posthumous *Fundamental Theory*, published ten years later. It is an appropriate time to attempt this, for nearly half a century has elapsed since Eddington's death. There is also a more important reason for this book. The ideas that Eddington thought were behind his books – and it will become clear to what extent these are truly the ideas behind them – were addressed to one specific problem. In the first half of the twentieth century, and certainly from 1926 onwards, physics became rather absurdly divided into two disparate parts. By 1916 general relativity had built a theory of gravitation on top of the successes of the special theory of 1905. It was highly successful on the large scale, notably in astronomy. This whole way of thinking about the world had nothing in common with that of the quantum theory, one form of which started at the turn of the century. Despite its crudity it had considerable success in explaining small scale phenomena. 1926 saw its replacement by a more refined approach but still one totally at variance with that of general relativity. Almost every assumption behind one approach was inconsistent with the other. Eddington was not the only scientist to be concerned about this ridiculous situation. As time has gone on, however, and physics has remained in the same unsatisfactory state, the consensus has been that it is almost indecent to mention the fact. The importance of another look at Eddington's ideas is that it serves to challenge the consensus.

Both of Eddington's books are of a technical nature, ostensibly addressed to the community of physicists, but here I am addressing the general scientifically-interested public, because the mystery is one of general interest. Accordingly, this is not intended to be a technical book. It tries to answer two questions: 'Why did Eddington come to hold the views that he did?' at both a historical and personal level and 'Were these views correct?'. The second has a technical component, but is truly a philosophical one. The problem is to unravel the various historical, personal and technical strands that are intertwined. At the same time I have to satisfy the community of physicists on some matters. The material for this is to be found in the numbered notes at the end of each chapter, and the corresponding number in the text shows the relevant place.[1]

The mystery

The mystery, in more detail, is this: Eddington was an outstanding astronomer of his generation and his *The Internal Constitution of the Stars* (Eddington 1926)[†] was definitive. Before that his *Mathematical Theory of Relativity* was instrumental in introducing the general theory to an English-speaking readership. The limpid style and elegance of these two books make them still a joy to read. Eddington's prose style is very distinctive. I am aware of only one comparable stylist of this century, and one whose style is remarkably similar, the economist Sir John Hicks. As a result of his later writings on science and philosophy Eddington became a household name in the 1930s. To the many scientific honours heaped upon him during his life was added the Order of Merit in 1938.

Everything changed, and not only for Eddington, in 1928 when Dirac published his wave equation for the electron, which was consistent with relativity and yet of a different form from any envisaged by the relativists. The new initiative prompted by Dirac's equation was what Eddington needed to make, as he thought, a break-through. Instead of a unified theory which would embrace both relativity and quantum mechanics, there was to be a bridge between them. The bridge would consist of certain problems having a particular kind of simplicity, so that they could be treated by either method. As we shall see later, such an idea depends on the truth of an unusual view of scientific theorising which Eddington had come to hold well before 1928. In the same year he began to publish papers on a new

[†] References are to be found at the end of the book.

initiative in physics, which he collected together in the first of the two books (Eddington 1936). This book was misunderstood and not very well received. Further papers followed and the other book (Eddington 1946) was more or less finished by his death and published posthumously by Sir Edmund Whittaker.

It cannot be denied that the arguments in both *Relativity Theory of Protons and Electrons* and *Fundamental Theory* are very obscure. This intense obscurity, following on from the clarity of the earlier books, is the essence of the Eddington mystery. How did he come to write such books? Was it simply that his powers of reasoning had become confused? That is oversimple. But E. A. Milne reviewed *Fundamental Theory* for *Nature* (Milne 1947). He had been an adversary of Eddington in the arguments over cosmology and relativity but he says, very fairly,

It must be recognised that Eddington has invented a mathematical technique, a logic of reasoning and a language of formulation which are as yet strange to most of us, and which no doubt conceal arguments and considerations which would be accepted if less individually expressed.

But there is a sting in the tail when he goes on to say:

whether or not it will survive as a great scientific work, it is certainly a notable work of art.

That the comment is meant to be critical is underlined by Milne's analysis of Eddington's claim of a second independent derivation of what he calls 'The central formula of unified theory':

Unfortunately for Eddington's aphorism, it emerges with the wrong power of N, in fact with an extra unwanted factor $N^{\frac{1}{3}}$, which makes nonsense of the so-called independent derivation. But Eddington is not one whit abashed. He calmly puts $\mu_1 = \mu_0(\frac{3}{4}N)^{\frac{1}{3}}$ . . .. But when Eddington had the bit between his teeth, no mere mathematical inconsistency held him back.

Today the majority opinion would probably be that Milne was being unduly generous. Even if you hold Milne's view, the question of why Eddington wrote *Fundamental Theory* is still relevant. Its answer would be found in psychology perhaps, but it would solve the mystery. On the other hand, if the ideas behind the two books are worth something, as many people still think, can these ideas be brought out into the open? This would dispel the mystery. The two approaches are not mutually exclusive because the matter is more complicated than either suggests. It is this that makes this present book worthwhile.

Relativity and quantum mechanics

Something should be said here about Eddington's attitudes to these two main theories of theoretical physics in the twentieth century and also about the criticism of his notion of a bridge between them. In both *The Internal Constitution of the Stars* and in *Mathematical Theory of Relativity* Eddington shows his awareness of the simultaneous developments taking place in quantum theory but he takes no part in them. This is particularly striking in the discussion of stellar structure. All the evidence for the discussion of the interior of stars comes from the radiation streaming out of them. This radiation is generated somewhere in the star and it was already clear in the 1920s that some kind of thermonuclear process was involved. It was for this reason that Eddington's great rival in astrophysics, Sir James Jeans, believed that it was too early to expect to understand stellar structure. Eddington's success was to see that the appearance of quantum phenomena could to a large extent be circumvented in the theory. Some problems of stellar evolution would no doubt continue to depend critically on nuclear theory and so became answerable only in the second half of the century, but the main lines could be made clear. The problem in general relativity was less serious. Eddington makes various references in his book to the curious features of quantum laws. On the whole, though, he is content to accept the 'old quantum theory' of before 1926 as a collection of rules of remarkable success since they were understood by nobody. When these rules were replaced by a more extensive theory, the 'new quantum theory' of 1925–6, Eddington saw that some of the new ideas might well be important in general relativity but he did not try to introduce them, for he did not yet see how.

It was not till the initiative inspired by Dirac's equation that Eddington had the notion of a bridge between the theories. Comparing treatments would give numerical values for certain physical constants. Amongst those claimed as a result of calculation by Eddington were the ratio of the masses of the proton and the electron (about 1836) and a constant that measures the strength of the electromagnetic interaction in quantum field theory, known (for historical reasons) as the fine-structure constant (about 1/137). The critical comments have tended to concentrate on this notion of calculating numbers that were otherwise thought of as known only experimentally. Eddington is partly to blame for this emphasis. But the comments have tended to ridicule any such theory as impossible and it is hard to see why this should be so. For, firstly, the idea of theoretical physics is to calculate numbers which will then be found by experiment, starting with numbers known already. The difference in Eddington's case was only that the empirical assumptions seemed relatively few.

Secondly, there is a well-accepted example of such a derivation of a numerical constant in electromagnetism. The theories of electricity and magnetism arose separately and each defined appropriate systems of units. Then Maxwell's theory completed the process of unifying the two. But the two sets of units, which had arisen historically, were not the same. A scaling constant was needed to reduce one set of units to the other and it had the dimensions of velocity and a numerical value close to the speed of light. Maxwell's theory predicted the propagation of waves with that speed. The obvious conclusion was that light was one example of electromagnetic radiation. Put it this way: there were two empirically-determined constants, both of the units of velocity. One was the scaling constant between electric and magnetic units; the other was the speed of light. Their ratio was very near to unity and Maxwell's theory then showed it to be exactly unity. Indeed, there is a much earlier analogy to such a calculation, noticed by Sir Edmund Whittaker. That is Archimedes' determination of limits on the value of π by determining the lengths of inscribed and circumscribed polygons. One could see π as a (very basic) physical constant whose value could be determined empirically by rolling a circular disc along a line. Archimedes showed that its value must lie between 22/7 and 223/71.

A simpler calculation in a more modern spirit than that of Archimedes is provided by considering squares inscribed and circumscribed to a circle of unit diameter. The perimeter of the circumscribing square is then evidently 4 and that of the inscribed square is $4 \times (\frac{1}{2}\sqrt{2}) = 2.828$ . . . .. These provide (rather crude) limits to the possible value of π and their average, 3.414 . . . is an approximation correct to about $8\frac{1}{2}\%$.

Thirdly, such an orthodox physicist as Dirac expressed the opinion that quantum field theory was incomplete precisely because it failed to calculate the value of the constant measuring the strength of electromagnetic interaction, the fine-structure constant, α. It is true that his argument was specific to this one constant. It rested on the fact that the whole procedure of quantum electrodynamics is meaningful only because the value of α is small enough. But even so, this is sufficient to show that the possibility of the calculation of some constants is not to be rejected out of hand. One can only regret that the criticism tended to concentrate on the existence of the calculation of numbers rather than on the obscure arguments by which the numbers were claimed to be found.

Sources for this book

The primary sources for this book are straightforward to list. The published papers are readily available but, in fact, largely unnecessary since the two final books contain the relevant material; the papers are useful only in dating ideas. Eddington also wrote extensively on the philosophy of physics and, in particular, on the view of physics implied by his later work. This writing has to be treated with some caution as it has a tendency to wish-fulfilment. But his *Philosophy of Physical Science* (Eddington 1939) cannot be ignored.

A skirmish previous to my own on the same field was carried out by N. B. Slater in his *Development and Meaning of Eddington's Fundamental Theory* (Slater 1957). Slater had available a large mass of manuscript material which proved to be earlier, more or less complete, drafts of *Fundamental Theory*. He collated these so as to try to see, from the genesis of difficult ideas, how they might be better understood. It will surprise no-one who remembers Noel Slater to be told that this is a major work of scholarship. After his death I acquired his photocopy of the manuscripts. I have been unable to derive any benefit from them which is not already set out in Slater's book which remains for me a convenient substitute for a primary source.

In addition I have made some use of a variety of secondary sources. Eddington's own life was not a very eventful one outside his work, but some details of it illuminate my investigation and here I have leant heavily on *Arthur Stanley Eddington* (Douglas 1956). But it is only fair to say that her interest is very much that of an astrophysicist, for whom the later work is obscure. Amongst philosophical discussions of Eddington, I have had considerable help from *Philosophie et Théorie Physique chez Eddington* (Merleau-Ponty, 1965). I also owe a debt to two of my former research students. Firstly, to Brian Tupper, with whom I published *Eddington's Statistical Theory* (Kilmister and Tupper 1962). This book grew from his PhD thesis, in which he made a serious attempt to reconstruct part of *Fundamental Theory* on a wholly rational basis. My present approach is different. For reasons that will become clear in the book, the most promising route to Eddington's ideas now seems to me to be the 1936 book, and, in particular, Chapters 11, 12 and 15 of it. I owe this insight to the second student mentioned, Michael Dampier, who in his 1969 thesis (Dampier 1969) explained that the apparently new discoveries of *Fundamental Theory* were, to a large extent, only developments of earlier results in

Relativity Theory of Protons and Electrons, with the aid of results from conventional physics.

The plan of this book is as follows. The nature of the questions posed means that they can be answered only by bringing together three strands. These are, firstly, aspects of Eddington's intellectual development, secondly, the structure and historical development of physics in the early twentieth century as it affected Eddington and thirdly, and perhaps to a lesser degree, the political and social circumstances of the time.

The book is divided chronologically into three parts: Part 1 begins with Eddington's early life and sets the scene by describing his intellectual development up to 1928. It is certain that an important change in Eddington's views took place when he tried to assimilate the consequences of Dirac's equation for the electron of that year, when Eddington was 45. Part 2 covers the six years from then till the ideas that were to lead to the publication of *Relativity Theory of Protons and Electrons* became clear to Eddington and Part 3 takes the story on over the next ten years till Eddington's death and the posthumous publication of *Fundamental Theory*.

Part 1 begins with Chapter 2 dealing with Eddington's education, his early work for the Royal Greenwich Observatory and his publication of *The Internal Constitution of the Stars*. The contents of this book are not directly relevant to our problems but the enthusiastic reception it received had important consequences. Eddington gained the respect of his peers and to this was added recognition by the population as a whole for his part in testing the general theory of relativity in 1919. His own thinking had been crystallised in a crucial way by the advent of general relativity in 1915 and so the development of this theory is described in Chapter 3 in relation to Eddington's life. This prepares the way for the discussion in Chapter 4 of the intellectual consequences of the theory for Eddington's philosophy. There is no way of finding out whether that collection of highly unusual views, partly expressed only implicitly, was formulated by Eddington only as a result of general relativity or whether he held some already, but the *Mathematical Theory of Relativity* contains the earliest exposition of them. In the concluding chapters of Part 1 I look forward to the coming Dirac equation in two ways, first in Chapter 5 by setting out in detail exactly what were the technical beliefs of Eddington which were so upset by it. Then in Chapter 6 I give a quick historical sketch of those parts of quantum mechanics which are needed to understand Dirac's innovation, which itself concludes Chapter 6.

Part 2 begins by discussing why and how Eddington came to the conclusion that non-commuting algebras were the key to Dirac's achievement and why his investigation of this led him to conceive the possibility of deriving by algebraic means the fine-structure constant and possibly also the three-dimensionality of space. Chapters 8 and 9 then analyse in detail two of Eddington's arguments, one for the fine-structure constant (he argues that $\hbar/e^2 = 137$) and one for the proton–electron mass ratio (his result is $m_p/p_e = 1848$).

Part 3 begins by assessing Eddington's state of mind in 1933, which led to his decision to start preparing *Relativity Theory of Protons and Electrons* for publication. Then in Chapter 11 I go forward to 1936–7 to deal with the adverse comments that the publication elicited and to begin to answer the questions, now in a more precise form, which were raised at the beginning of this book. Finally I conclude in Chapter 12 with the writing of *Fundamental Theory*, the reasons for it and the extent to which, notwithstanding appearances, it failed to meet the criticisms that had been urged against the earlier book, or to go significantly beyond it.

For convenience, I shall abbreviate the names of the two books under discussion to RTPE and FT respectively.

Notes

1. I should say something here about technical notation, and my departures from Eddington's. Writing in 1923, Eddington used the letter B for the Riemann–Christoffel (RC) tensor; R is now universal, so I have used it. He also used G for its contracted form, the Ricci tensor, which is also universally written as R (and merely distinguished from the other by the number of suffixes) and again I have conformed to modern usage. So as not to depart too far from Eddington's algebraic usage, it is essential to define the special relativity metric, for which I use η, as $\eta_{11} = \eta_{22} = \eta_{33} = -1, \eta_{44} = 1$, and this I have done. But my major departure from Eddington is in the definition of the Clifford algebra. Eddington defined his 'single frame' as generated by four anti-commuting square roots of -1. To relate this to special relativity requires the use of imaginary time and all the bad old ways introduced by Minkowski. Instead, I have followed modern usage in defining four E_a, to generate the algebra, by

$$E_a E_b + E_b E_a = 2\eta_{ab}.$$

No confusion should result, as I use Latin suffixes in place of Eddington's Greek. These notations and others will be fully explained when they are introduced but are mentioned here to prevent confusion when reference is made to Eddington's books.

PART 1

1882–1928

2

The astrophysicist

It will be clear from Chapter 1 that the story of Eddington's intellectual life cannot be dissociated from that of the fascinating revolution of ideas in physics between 1900 and 1930. I shall not try to do so, but my history of the general scientific ideas will be a very partial one, directed to those aspects which bear in any way on Eddington's development. The first part of this chapter is concerned with personal features of Eddington's earlier life that I believe to be relevant in judging his later work. More detailed biographical detail is to be found in *Arthur Stanley Eddington* (Douglas 1956) on which I have relied. The historical context of the advent of special relativity comes next, and finally I deal with Eddington's definitive account of stellar structure as the rounding off of his earliest preoccupations.

Eddington's early life

Eddington was born in Kendal on 28 December 1882, into a Quaker family, and his faith played an important part throughout his life. He showed evidence from an early age of a prodigious memory and an interest in very large numbers. The family removed to Somerset after his father's death and Eddington received a Somerset County Council Scholarship in 1896 to go to Owen's College, Manchester though he was still under sixteen. His four year course there began with a general year, and then followed three years of physics (under Schuster) and mathematics (under Lamb). Lamb provided more than mathematics; his prose style was the model for

Eddington's own carefully nurtured style. In common with other similar university institutions of the time, the syllabuses of the mathematics course were constructed to mirror what was seen as the best model of the subject – the Cambridge Mathematical Tripos of some considerable time earlier. Thus Eddington was not taught the modern mathematics of the turn of the century but he did learn to solve very difficult problems. To a lesser extent the same is true of the physics, but one can well imagine under Schuster that a good training in practical physics was considered essential.

After receiving his first class BSc in physics, Eddington went up to Cambridge, to Trinity, in 1902 to read for the Mathematics Tripos. Such a course of action was not uncommon at the time, and for many years after, for graduates of the increasing number of red-brick universities, as well as from the ancient universities of Scotland. He took Part I after two years and was placed Senior Wrangler, that is, the first in the list in the old Tripos in which the Pass Lists gave the candidates in order of merit. It was, above all, a test of endurance, not very much changed over the previous century from the form in which it had evolved by slow stages from the 'Senate House Examination'. This had itself evolved in the eighteenth century as a replacement for the medieval disputations. Indeed, Eddington, himself, would in the future serve on the committee charged with reforming the Tripos. The examination consisted of seven papers sat over four days, on more elementary material, followed, a week later, by seven more advanced papers, again over four days. Euclid still played an important part in the first four days and Newtonian mechanics in the second, though more variety had been introduced by 1904. So here, though somewhat more modern than Eddington's Manchester teaching, the emphasis was still on very hard problems. The whole Cambridge view on mathematics was beginning to change, under such influences as A. N. Whitehead, but the situation was not yet much different from that noted by Grace Chisholm some ten years earlier (Grattan–Guinness 1977):

Mathematical science had reached the acme of perfection. Through the long future ages, no new ideas, no new methods, no new subjects were to appear. The edifice of mathematical science was complete, roof on and everything . . ..
  Everything pointed to examinations, everything was judged by examination standards, progress stopped at the Tripos.

Russell at about the same time made similar criticisms (Russell, 1967):

My mathematical tutors had never shown me any reason to suppose the Calculus anything but a tissue of fallacies . . . .. The mathematical question [of two that worried Russell] had already in the main been solved on the Continent, though in England the Continental work was little known. It was only after I left Cambridge and began to live abroad that I discovered what I ought to have been taught during my three years as an undergraduate.

That describes the position in pure mathematics. In applied mathematics some changes were made but the emphasis on Newton as the basis for everything was still strong.

This background is useful in understanding the attitude that Eddington's early mathematical training produced in him. In the early 1930s, for example, he declined to consult texts on modern algebra to assist him in his new developments, saying 'I like to work these things out for myself.' During his studies for Part I, in November 1903, he had also taken the University of London's External BSc general degree in mathematics and physics. Here again, such 'test-runs' were not all that uncommon. The syllabus in London was more on the Owen's College lines; he was placed as a first in mathematics and a third in physics. He stayed on in Cambridge to take Part II of the Tripos in 1905 and the autumn of that year found him in Cambridge again doing some part-time tutoring.

But early in 1906 the Astronomer-Royal offered him the Chief Assistant post at the Greenwich Observatory, in succession to Dyson (who had gone to Edinburgh). Eddington agreed and was appointed early in 1906. His ties to Cambridge were not severed at once. He was awarded the Smith's Prize in 1907 and elected a fellow of Trinity. Whilst at Greenwich he lived in Blackheath Village, at 4 Bennett Park; the house is now marked with a plaque. From there it is a pleasant walk through the village, up onto the Heath and so on through Greenwich Park to the observatory. His duties there were largely practical, but he maintained an active theoretical interest too. When the Plumian Chair of Astronomy fell vacant, he was a fairly obvious choice and so by 1913 we find him, at the age of thirty, back in Cambridge in the post which he was to occupy for the rest of his life. In 1914, on the death of Sir Robert Ball, he became Director of the Observatory as well. He was elected to the Royal Society in the same year.

Science at the turn of the century

The decade from Eddington taking Part II had been one of turmoil in mathematical physics and I want now to consider this context in which he was working. The turn of the century had seen the introduction of the so-called old quantum theory which proclaimed in an *ad hoc* way that certain physical quantities associated with atomic phenomena, such as energy, were not continuous, as had always been supposed, but came in *quanta*, that is, small packets of a common size. Two different kinds of consequence follow. In the first place, the resultant theoretical system is no longer a consistent one. The framework into which the discontinuous energy was to be fitted was one constructed on the basis of the continuity of energy. The inconsistency did not worry physicists greatly at the time. It was only in the early 1920s, when the old quantum theory found itself embedded in its own internal contradictions, without knowing how to proceed, that something had to be done. Secondly, atomicity became seen as a phenomenon applying more widely than had been believed at any time since Lucretius. The importance of atomicity affected Eddington profoundly. I will have more to say about the influence of the actual quantum theory on Eddington; this is a complex topic. At this stage his assimilation of it was less important than that of special relativity, the theory which arose in the paper *On the Electrodynamics of Moving Bodies* (Einstein 1905).

Einstein couched his argument in terms of a problem in electrodynamics. In modern terms one could say he was motivated by the problem that Maxwell's equations for the electromagnetic field appeared to take a different form for two observers in uniform motion, one relative to the other.[1] The notion of an 'observer' is widely used in explanations of special relativity. No human observer is necessarily intended. One can always, with more or less circumlocution, avoid such anthropomorphic language if one introduces instead the notion of a (spatio-temporal) reference-frame. This notion is a generalisation of that of a cartesian coordinate-system (Fig. 2.1). It is generalised in two different ways. Consider first the case of plane coordinates. If the two axes are not at right-angles it is usual to distinguish between the *contravariant* coordinates, where the distances to the point are measured parallel to the axes, and the *covariant* coordinates, where perpendiculars are let fall on the axes and the intercepts that they make are taken as the coordinates. Secondly, recall that in physics one is concerned with *events*, that is, points associated with instants of time. A reference-frame is a set of rules for assigning to an event three spatial coordinates and then one more number to determine its time of occurrence.

The difference in form of Maxwell's equations for two reference frames in uniform relative motion is contrary to the expectations aroused by Newtonian mechanics, in which two such observers would find the same dynamical equations.

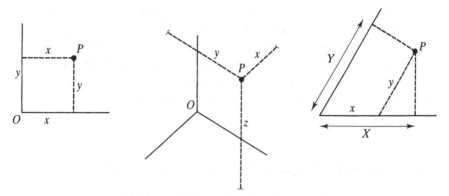

Fig. 2.1. Methods for specifying the position of P.

Einstein was joining in the same general discussion as Poincaré in France and Lorentz in Holland. But whereas they had both been primarily concerned with the need to clarify the mathematical details of electromagnetic theory, Einstein's paper tackled the problem at a more fundamental level, and one at which electrodynamics played only a minor role. He drew attention to the following hitherto unremarked fact. The notion of the time at which events occur was a clear notion and well defined for events near to an observer, so long as he is equipped with some kind of clock. The same could not be said when the events were distant from the observer. A time can be assigned to such events only by means of signals sent between event and observer. This sounds like a practical view and Einstein does indeed say 'We arrive at a much more practical determination by going along the following line of thought.' He then introduces the signal idea and takes the signals to be electromagnetic (light), which is the remaining minor role allotted to electromagnetism in his theory.

He goes on to say 'The latter time [that to be assigned to the distant event] can now be defined by establishing *by definition* that the "time" required by light to travel from A to B equals the "time" it requires to travel from B to A.' Notice that, in keeping with the complex interaction between observation, convention and theory almost always to be found in Einstein's thought, the time is assigned by definition. That signals will need to be used and that light is a convenient signal are commonplaces. But before 1905 it had been supposed that the signals were simply the means used to determine the time, which was unique and which existed independently of them. This Einstein denies and his argument has the serious consequence that the universal time of Newtonian mechanics has to be given up. The time of an event depends both on the event and on the reference frame by which it is measured. As a result, the equations for getting the coordinates and time in one reference frame from those in another in uniform motion relative to it are changed. Precise details of how this change results are given in the notes[2] but for present purposes it is sufficient to note the results. If the direction of motion is taken as the x-axis, the 'Galilean transformation' of Newtonian theory is

$$x \to x' = x - Vt,$$

and this has to be changed to the Lorentz transformation:[2]

$$x \to x' = \beta(x - Vt), \ t \to t' = \beta(t - Vx/c^2)$$

where

$$\beta = \frac{1}{\sqrt{(1 - V^2/c^2)}}.$$

This serves to put right the problems about Maxwell's equations. But the price that has to be paid for this is that of a major change in ontology. The Newtonian world had a concept of time as a number assigned to an event. This has to be replaced with the assigning of a number to a combination of an event and a reference-frame.

The Cambridge view

Such is the real content of Einstein's paper but this is not quite how things looked when seen from Cambridge. The influence of Maxwell himself had been profound. He had inferred from his electromagnetic equations the existence of wave solutions travelling with the speed of light and he had identified light itself with such a solution. The problem for Maxwell, and many British physicists, was the medium of transmission of the waves. This supposed medium, the aether, was seen by Maxwell as a physically real diffuse medium filling the universe. The definitive account of this at the turn of the century was *Aether and Matter* (Larmor 1900). In it, aether theory had assumed a very subtle and elegant form. But if the aether was physical it should be possible to determine the motion of the earth through it by optical experiments. Maxwell suggested such an experiment, which was carried out in America by Michelson, first alone and then in conjunction with Morley.

Fig. 2.2. The Michelson–Morley experiment.

In the Michelson–Morley experiment a beam of light is split in two by a half-silvered mirror. The two parts travel out and back along two directions at right-angles, being reflected by mirrors at the ends of two rigid arms (Fig. 2.2). The relative time of the return rays is monitored by means of optical 'interference'. A pattern of light and dark bands is obtained on a screen, because the two paths do not differ by an exact number of wavelengths. The apparatus is then rotated in the same horizontal plane through a right-angle. If one arm were, for instance, initially moving forward through the aether, so that the light speed would be different from that along the other transverse arm, then after rotation the situation will be reversed. The first arm will now be transverse to the 'aether wind' and the other arm will be suffering the effect of opposite movement. The interference fringes will therefore shift. No such effect was found, nor was this a coincidence resulting from the earth just happening to be at rest in the aether at the time. For a repetition six months later again gave a null result.

Einstein makes no reference to the Michelson–Morley experiment in his paper and at one time believed himself to have been ignorant of it, but later he admitted that it must have had some unconscious effect on his thinking. In the very much more aether-conscious atmosphere of British physics, the experiment was seen as an empirical demonstration that it was impossible to detect the motion of the earth through the aether. This led to a variety of responses. Few were yet happy to draw the obvious conclusion that, unfortunate as it might be for any intuitive need to have a medium in which light was travelling, no such aether existed. Some found Einstein's explanation of the Michelson–Morley experiment (that, as a simple consequence of the Lorentz transformation, the speed of light, unlike all other speeds, is unchanged by the motion of the observer) uncongenial and preferred to explain the result by hypothesising changes in the lengths of the arms of the apparatus. At the extreme of this party can, perhaps, be placed Rutherford who, as late as 1910 rejoined, to the suggestion that no Anglo-Saxon could understand relativity, 'No! they have too much sense'. Between the extremes were those who accepted Einstein's theory with greater or less enthusiasm, and saw as the next problem the rendering of it and aether theory as a consistent whole.

Eddington on relativity

Eddington was not one of the early protagonists in this discussion. He showed little interest in special relativity until the advent of the general theory in 1916. Then he began as one of the more enthusiastic members of the centre party.

As late as 1920 (Eddington 1920a) he explains the contraction of the arms of the apparatus on these lines:

When we consider the matter carefully, it is not so surprising after all. The size and shape of the material apparatus is maintained by the forces of cohesion, which are presumably of an electrical nature, and have their seat in the aether. It will not be a matter of indifference how the aether is streaming past . . .. But what does seem surprising is that the readjustment of size should just hide the effect we were hoping to find. It almost looks like a conspiracy . . . the conspiracy is a general one; and, in fact, it is in the nature of things that this motion is undetectable.

We make this generalisation and build a branch of science on it, in the same way as we make the generalisation that it is impossible to construct a perpetual-motion machine, and build the science of thermodynamics upon it.

But although he began from such a firm aether standpoint, Eddington came to a different understanding in the next three years. He realised that the essential contribution of Einstein was to lift the controversy from the narrow realm of electromagnetism to the broader one of time-reckoning. This allowed Eddington to make a lasting contribution. In the *Mathematical Theory of Relativity* (Eddington 1923) he considers a different attack on the problem raised by Einstein, that of correlating time-measurements at different places. Eddington imagines a set of clocks, initially all synchronised at one place, and then moved about, before the experiments start as it were, to different places, but very slowly, so slowly that any effect of motion on the clock-running is negligible. Then he says (p. 15):

I do not know how far the reader will be prepared to accept the condition that it must be possible to correlate the times at different places by moving a clock from one to the other with infinitesimal velocity. The method employed in accurate work is to send an electromagnetic signal from one to the other, and we shall see in §11 that this leads to the same formulae.

Then he goes on, in a passage which is important in making clear how his view differs from Einstein's clear statement that the distant time is just formulated *by definition*. Eddington prefers to say:

We can scarcely consider that either of these methods of comparing time at different places is an essential part of our primitive notion of time in the same way that measurement at one place by a cyclic mechanism is; therefore they are best regarded as conventional.

It is only fair to mention that he goes on to point out that this convention agrees with empirical practice.

Some eight pages later he feels the need to expand this conventionalist standpoint so as to bring out its true complexity and subtlety better. There he says:

Much confusion has arisen from a failure to realise that time as currently used in physics and astronomy deviates widely from the time recognised by the primitive time-sense . . .. Our time-sense is not concerned with events outside our brains; it relates only to the linear chain of events along our own track through the world .

The external events which we see appear to fall into our own local time-succession; but in reality it is not the events themselves, but the sense-impressions to which they indirectly give rise, which take place in the time-succession of our consciousness . . . the idea has arisen that the instants of which we are conscious extend so as to include external events and are world-wide . . .. This crude view was disproved in 1675 by Römer's celebrated discussion of the eclipses of Jupiter's satellites . . .. The whole foundation of the idea of world-wide instants was destroyed 250 years ago, and it seems strange that it should still survive in current physics. But, as so often happens, the theory was patched up although its original *raison d'être* had vanished. Obsessed with the idea that the external events had to be put somehow into the instants of our private consciousness, the physicist succeeded in removing the pressing difficulties by placing them not in the instant of visual perception but in a suitably preceding instant . . .. We need have no quarrel with this very useful construction which gives physical time. We only insist that its artificial nature should be recognised, and that the original demand for a *world-wide* time arose through a mistake . . .. It is important for us to discover the exact properties of physical time; but these properties were put into it by the astronomers who invented it.

This last sentence is the first time in his writings that Eddington formulates an idea which was to recur frequently. At least some of the 'properties of the external world' found experimentally by physics are there as a result of the theoretical and experimental conventions involved in their measurement. And it is only a small step (though not one which Eddington was yet prepared to take) to form a corollary to that view. If such properties are there for that reason, they might be found by an alternative route, by the careful analysis of the means of observation. Eddington was initially pushed into this position by the impact of relativity. It is the key to understanding much in his later writing. He expounded the idea in various places (e.g. Eddington 1939) under the name of *selective subjectivism*, and it is now widely associated with his name, mostly with disapproval.

But earlier in the same passage occurs another characteristic Eddington idea that has not been widely recognised at all. I shall call it *descriptive tolerance*. It is captured by the phrase 'We need have no quarrel with this useful construction.' Eddington took for granted the idea that scientific concepts were no different from everyday ones in the way in which one could use them, or not, as convenient. Physical time, for example, is not 'out there'; it is the result of a construction, an idea to be used if desired. It is not compulsory to talk in a particular way. I believe this to be quite at variance with the usual attitude of scientists. To them, physical time is 'correct' and the everyday notion is an incorrect approximation, which depends for its validity on being an approximation to the correct. Eddington would say, on the other hand, that the community of physicists uses certain languages. (In these post-Wittgenstein days one would say: takes part in certain language-games.) Usually the choice of language is obvious but not always. Descriptive tolerance is the notion referred to in Chapter 1 as a necessity for the validity of Eddington's idea of a bridge between theories constituted by simple problems capable of description by each of the two.

These two ideas are the beginning of what may be called the later Eddington. They developed farther and were joined by others after 1916 when the general theory of relativity was published. But the way in which they developed requires a separate chapter and so I break into the chronological order of Eddington's life at this point and consider the closing phase of the earlier Eddington, the astrophysicist.

Astrophysics

From 1906 onwards Eddington had published a series of astronomical papers, initially of a practical variety, but becoming more theoretical, until in 1916–18 came a series of four papers 'On the radiative equilibrium of the stars' (Eddington 1916) which marked the beginning of a new phase of work. This work was summed up in a book (Eddington, 1926), a definitive account of its subject. Eddington was led into this work by his attempt to discuss the mechanisms of Cepheid variable stars (Eddington, 1917, a paper delayed in printing). These are stars with a characteristic variation in brightness. Those near enough to the earth to have their distances measured by direct astronomical means are found to have their intrinsic brightnesses (the observed brightness corrected to allow for the distance away) dependent in a definite way on their periods of variation. If it is assumed that all the other Cepheids have the same property, the fainter ones will serve as a means of determining their distances. (It now seems that there are two kinds of Cepheid variable, not one, so that the distance scales have had to be revised, but that is a later development.) The importance of the Cepheids as an astronomical tool led Eddington to wonder how they worked, how the mechanical energy of pulsation was maintained against dissipative loss. Before he could discuss such oscillations he had first to revise the existing treatments of the stationary state.

He begins by admitting that the interior of the sun and of stars seems an inaccessible region. But, he argues, this is really far from the case. The interior communicates with us in two ways: through its gravitational field and through the radiation which flows out across the surface. Then he assumes that these are the important controlling factors. Such an assumption represents a streak in Eddington's thought which recurs. Since these are the only observable parameters, it is evident to him that the right course of action is to begin with a theory using them as controls. Only if that theory is falsified is it necessary to look further. I know of no evidence of any direct influences between Popper and Eddington, but the spirit is the same. And there is another idea at work here, which was to develop in Eddington's later investigations – that of physics as structure. The understanding of the whole nature of the star might well be impossible at the time but the structural features exhibited by the gravitational field and the radiation are sufficient to determine something. It turns out that this structural physics of the stars is adequate to fix a great deal.

In common with earlier workers, Eddington sees the star as having an outer atmosphere whose physical properties depend on these two influences. The star stays in a steady state (for most of its life) because the atmosphere adjusts to let out the radiant heat. This means that the surface conditions are determined by two parameters, g, the gravitational field at the surface, and T, the effective temperature. One therefore expects a two-fold sequence of stellar spectra. From 1914 this was accepted as so; to the Draper sequence of classification by types (which is roughly classification by T) was added the classification by absolute magnitude (which turns out to be according to g). But there is another way of classifying, if we look at the star as a whole, and that is according to the mass, M, and the radius, R. How are these two different classifications to be related? In particular, how is T, which determines the flux of radiation, determined by M and R?

Nernst and Jeans had argued that this problem was insoluble. It was already clear that the heat was produced in the interior by some kind of thermonuclear process; they pointed out (in the 1920s) that the laws of nuclear physics were almost completely unknown. Eddington saw that another approach was possible:

The amount of water supplied to a town is the amount pumped at the water-works; but it does not follow that a calculation based on the height of the water and the diameter of the mains is fallacious because it avoids the problems of the pumping station.

The temperature in a star (a sphere of gas obeying the simple gas law perfectly) was investigated in the nineteenth century (Lane 1870). The problem Lane formulated was completely solved by Emden in his *Gaskugeln* (Emden 1907). Lane found that the temperature of a contracting star rises. His theory was ruled out by the general belief that young stars are hot and old ones cold. In 1913 Hertzsprung and Russell put forward the giant/dwarf theory, in which stars start as low density, high mass, cool red stars. They contract and so increase in temperature until the perfect gas law no longer holds on account of the high density. The star then cools again. In the ascending series of giants the greater surface area produces greater luminosity and so greater (visual) magnitudes.

Eddington's approach to the theory was characteristic. He accepted the ascending series but refused to give up the perfect gas assumption in the descending series of 'dwarfs'. To deny the assumption was to admit inability to deal with the problem, for no theory of the liquid or solid states was available. With another Popperian choice, Eddington retained the gas law and this gave a testable theory. It turned out to fit the observations well. It works with three variables, the pressure, the density and the temperature. These are related by the perfect gas law, so only two further equations are needed to formulate the problem. These are, respectively, the conditions for mechanical equilibrium and for thermal equilibrium. But these are not simple algebraic equations; they arise by looking at the equilibria of small sections of the star, and involve therefore the rates of change of the quantities. In the mathematician's language, they are differential equations to be integrated.

Eddington's outstanding contribution was to carry out this process in two special cases: (i) under the assumption that the energy was being generated uniformly per unit mass; (ii) the other extreme, using the 'point source model' in which all the energy was generated at the centre of the star. The second case was very much more difficult to carry out, for it requires laborious numerical calculations. Fifty years later a computer would have made these calculations readily, but as it was much time was consumed. This was in the nature of the problem:

The mass M and the radius R of the star are only found at the end of the calculation; in their place we must have two disposable initial conditions. Unfortunately we can only make a beginning by fixing *three* disposable constants . . .. We thus over-condition the problem and generally fail to reach a solution . . .. After many trials we contrive to straddle the true solution sufficiently closely. Of course, a great amount of calculation is wasted on unsuccessful trials.

The two extreme cases gave essentially consistent results, in accordance with the 'pipe-line' analogy. The result is the 'mass–luminosity relation' since the radius turns out to play no important part. It is then possible to calculate the central temperature. This comes out roughly to 4×10^7 K for all main sequence stars. Such a surprising result points to a mechanism of energy production which is extremely temperature-dependent.

Here was the adult triumph of a child's interest in large numbers: a child who set himself the task of counting the letters in the Bible, and succeeded in getting well into Genesis, and who had learnt the twenty-four times table before he learnt to read. Alongside this humdrum work he had begun excitedly to recast his philosophical ideas under the influence of general relativity, to which I turn in the next two chapters.

Notes

1. That is to say, if Maxwell's equations,

$$\text{curl } \mathbf{E} = -\dot{\mathbf{B}}, \qquad \text{div } \mathbf{D} = \rho,$$
$$\text{curl } \mathbf{H} = \dot{\mathbf{D}} + \mathbf{J}, \qquad \text{div } \mathbf{B} = 0$$

hold in one coordinate-system, then they fail to hold in a new system defined by the Galilean transformation from the first one:

$$\mathbf{r} \to \mathbf{r}' = \mathbf{r} - \mathbf{V}t.$$

Maxwell does not accept this failure but he has to avoid the difficulty in a particular situation in which a stationary charge e is under a force $e\mathbf{E}$ by remarking that in the new coordinate-system the charge has speed $\mathbf{V}$ and so constitutes a current $e\mathbf{V}$ which accordingly suffers an additional force $e\mathbf{V} \times \mathbf{H}$. Einstein has recorded how a more acute form of this non-invariance disturbed him from his youth. He considered a plane-wave solution to Maxwell's equations, that is, a light-wave, which is known to be transverse. If one transforms so as to be moving along with the wave, the result is a static planar solution of Maxwell's equations; but no such solution exists.

2. In an anachronistically modernised one-dimensional version of Einstein's argument for two inertial observers in uniform relative motion in the x-direction, the observer O *assigns* time $t = \frac{1}{2}(t_2 + t_1)$ and distance $x = \frac{1}{2}(t_2 - t_1)c$ to the event E in the space-time diagram (Fig. 2.3). Hence

$$t_2 = t + x/c, \; t_1 = t - x/c.$$

If the two observers synchronise their clocks at the moment when they are together, then clearly

$$t' - x'/c = k(t - x/c),$$

where k is some constant measuring the rate of separation. But, assuming the equivalence of all inertial observers, the returning signal times will satisfy

$$t + x/c = k(t' + x'/c)$$

with the same value of k. An immediate consequence is that

$$s^2 = t'^2 - x'^2/c^2 = t^2 - x^2/c^2$$

is the same for each observer, and solving the original equations gives the explicit relation between their space and time measurements in the form of the Lorentz transformation:

$$x' = \beta(x - Vt), \; t' = \beta(t - Vx/c^2)$$

where

$$\beta = (1 - V^2/c^2)^{-\frac{1}{2}} \text{ and } V = (k^2 - 1)/(k^2 + 1).$$

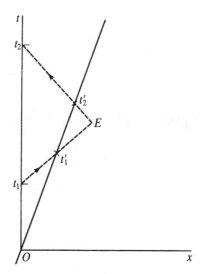

Fig. 2.3. Determination of time and distance of a distant event by light-signals.

3

General relativity

This chapter discusses the historical impact of the advent of general relativity, in 1916, on British physicists and astronomers, and, at a personal level, its effect on Eddington's life. In this way I prepare for the next chapter in which I shall show that the intellectual effect of the general theory on Eddington was to accentuate certain philosophical ideas which he already had. These ideas take us some way along the road to understand the mystery described in Chapter 1.

The genesis of general relativity

We left Eddington reinstalled in Cambridge in 1914, clearing up his earlier astronomical work and just about to begin the investigation directed towards the understanding of the mechanism of the Cepheid variables. By that time the dust was beginning to settle on Einstein's 1905 paper in which the concept of time had been so strikingly changed. The Cambridge aether-theorists were working their way towards reconciliation but, doubtless, were more concerned in 1914 with the international disaster that was to sweep away the comforts of the long Edwardian summer. Then in 1915–16 there appeared a new paper by Einstein (Einstein 1916) which followed up his earlier one by changing completely the concept of space. This is usually described as Einstein's successful attempt to extend the idea of relativity to include gravitation. The problem of gravitation was considered particularly important because special relativity seemed to give a privileged position to the other well-known field theory, electromagnetism. The Lorentz transformations left Maxwell's equations unchanged. Yet in some ways gravitation was the more fundamental field because of its universal action. There are no 'neutral bodies' in gravitation as there are in electromagnetic theory, no shielding of gravitational effects. Newton's gravitational theory was not left unchanged by the Lorentz transformations. Gravitation therefore presented itself as the next challenge.

Einstein's paper was sent to Eddington by de Sitter in neutral Holland, for access to German journals was cut off. Eddington was greatly impressed and lost no time in making a first explanation of it (Eddington 1918) to the scientific community. This showed, incidentally, that he had very quickly learnt the necessary mathematics (which seemed at the time very formidable). I shall set out in the next few pages the thoughts behind Einstein's paper (Einstein 1916), for Eddington seems to have grasped some of these, though they are not evident in the printed form.

The special theory of relativity was formulated in terms of the concept of an *inertial observer*, that is, one who observes as true Newton's first law of motion (often called on the Continent, the Law of Inertia). This states that bodies under no forces continue at rest or moving with uniform speed. But because of the way that the theory uses light rays as signals, the definition of inertial observer is restricted further for Einstein. The inertial observer must also observe Maxwell's electrodynamics as true, and so find light travelling in straight lines. The theory had resulted in a new concept of time. This, in turn, led to a variation in the relations holding between the measurements made by two inertial observers in uniform motion one relative to the other. They would no longer agree about the spatial distance PQ between two events at P and Q. This is because their measurements of PQ require by definition the *simultaneous* location of P and Q and the theory results in different criteria of simultaneity for the two observers. In the same way the lapse of time, T, between the events is measured differently by the two observers. But to compensate for this the theory does point out a quantity expressing the separation of the events about which the observers agree. This is the so-called proper time, S, given by $S^2 = T^2 - (PQ)^2/c^2$, where c is the speed of light. And although of course the two observers will agree about the truth of Maxwell's equations, the other classical field theory, gravitation, has been ignored in this whole discussion.

The first idea for dealing with this omission was to modify Newton's theory of gravitation to 'make it Lorentz invariant', that is, to replace it by a new theory, not too different, which assumed the same form for all inertial observers, just as Maxwell's theory did. It is, in fact, not very difficult to do this. The advantage of such a Lorentz invariant theory was that two inertial observers would then see the same gravitational field. This apparent advantage ignores an important aspect of gravitation, that a local gravitational field can be removed by an acceleration (as when a lift suddenly falls or an astronaut is 'weightless' in orbit). Einstein came to see this as paramount around 1907. Meanwhile several candidates appeared for Lorentz invariant theories. Their agreement with experiment was in doubt but this does not seem to have played an important part in Einstein's thinking. Instead it became more and more clear to him that the problem had not been correctly formulated. This he saw because a uniform Newtonian gravitational field (say, approximately, that near the surface of the earth) can be created or removed by an accelerated coordinate-transformation (as when a lift starts suddenly up or down).[1] Such a change in coordinates turns straight lines into curved ones (parabolas) so that, assuming that the non-uniform field round a massive body is of the same general kind, it follows that light will be slightly bent in passing near such a body.[2] The only practical way of observing such a small effect is by looking at the change in the apparent position of stars when the path of their light has grazed the sun's disc. Of course, in order to see the stars, this has to be done at the time of a solar eclipse.

It is only a gravitational field that can be created or destroyed by a coordinate-transformation in this way. This is because it is an *acceleration field*. Bodies of different masses at one point in the field fall with equal accelerations, a fact known already to Galileo. This 'principle of equivalence'[3] was recognised as important by Einstein in 1907. He saw that, although the problem was not, after all, to formulate a Lorentz invariant gravitational theory – that would have been appropriate only for inertial observers – this incorrect requirement pointed to something deeper that was needed. Essentially the Lorentz transformations are important because of the critical role played by light in signalling. This critical role is exhibited by the fact that the speed of light is the greatest allowed speed. No effects can be transmitted faster than this speed. Now this maximal quality of the speed of light is in conflict with Newton's inverse square law of gravitation, because that implies an instantaneous gravitational effect, and it is this conflict that needs to be resolved.

Einstein's path in the next few years was a tortuous one that need not concern us here. Suffice it to say that it took in a consideration of the rotating rigid disc (gramophone turntable). Though the theory of such a disc was beyond the powers of special relativity, it seemed clear that, because of the changes in time-reckoning implied by the Lorentz transformation, there should be an apparent contraction of the circumference of the disc, whereas the radius would be unchanged. So the apparent geometry on the disc would no longer be one in which the ratio of circumference of a circle to its diameter was π; that is, it would no longer be Euclidean. One consequence of this was that, whatever coordinate-system one employed for the geometry of the disc, these coordinates would not, in general, have their 'usual metrical meaning'. They would simply be labels for the points of the disc. In some way, then, gravitational fields, accelerated reference frames, non-Euclidean geometry and generalised coordinates were all connected. But it was still hopelessly unclear how the connexions were to be made.

Differential geometry enters

At this point Einstein returned to Zurich (August 1912) to consult with 'my friend, the mathematician, Marcel Grossmann'. Einstein generously thanks Grossmann in his published paper but does not make it quite clear what Grossmann's first contribution was. I am sure, from internal evidence, that it was to remind Einstein of the lecture course they had attended as students on Gauss's theory of surfaces. This was crucial for aiding Einstein in seeing two aspects of the connections posed by the rotating disc problem. Although this is not set out clearly in Einstein's paper, it was also crucial in Eddington's understanding of the theory. The first aspect was this: Gauss considered an arbitrary, smooth surface in ordinary three-dimensional space. A point on the surface is specified in terms of two parameters, like latitude and longitude on the surface of the earth. Gauss uses these parameters as coordinates for describing the geometry on the surface. So again, here are coordinates 'without their usual metrical meaning'. This metrical meaning only comes *via* the 'second quadratic form' which is the variant of Pythagoras' theorem for measurements on the surface.[4] Thus geographers are well aware that, at latitude L, the result of *small* changes in both latitude and longitude is to generate a displacement D on the sphere given by

$$D^2 = R^2[(\text{change of latitude})^2 + (\cos^2 L)(\text{change of longitude})^2],$$

where R is the radius of the earth. The quadratic expression in changes in coordinates, whose value is the square of the distance generated by the changes, is called the *metrical form* for the surface. The general case may have a product term as well as the two squares, so there are three coefficients. These three are called the metrical coefficients, or simply the *metric*.

This metrical form is a property of the surface which does not change if the surface is bent without stretching or tearing. One could call this an *intrinsic* property of the surface. It may be helpful to think of intrinsic properties in a more picturesque way. They are those which could be discovered by little two-dimensional creatures living on the surface and unable to get outside it. The second way in which the Gauss theory was essential for Einstein was in carrying through the distinction between intrinsic properties and other (extrinsic) ones to the description of the curvature of the surface.

One can determine mathematical expressions in different ways, any of which might well be called the local curvature of the surface. The different formulae correspond to different concepts of curvature. Most do not represent purely intrinsic properties. If a sheet of paper is bent smoothly, without creasing, into the form of a cylinder, the geometry on it is quite unchanged from when it is flat. Yet there is an obvious sense in which the surface is now curved whereas previously it was flat. The concept of curvature that describes its cylindrical shape cannot therefore be an intrinsic concept.

On the other hand the surface of a sphere has an intrinsic curvature for the angle sum of a triangle drawn on it exceeds 180° by an amount proportional to its area. Gauss was able to show that the curvature of the surface was completely measured by two numbers at each point, one of which was intrinsic. This intrinsic part of the curvature depended only on the metrical form and on how it varied from point to point.[5] It is now introduced by means of the idea of 'parallel displacement' to use the modern term. A direction A on the surface at a point, if displaced exactly as it is to a nearby point, will usually no longer be a direction on the surface (Fig. 3.1). But one can always split the displaced direction into a part N along the direction normal to the surface (that is, the direction at right-angles to the tangent-plane) and the other part B along the surface. This part on the surface is said to be derived from the original by parallel displacement.

As well as explaining all this, Grossmann was able to tell Einstein that the Italian differential geometers had generalised Gauss's treatment to any number of dimensions. The treatment of time in special relativity had

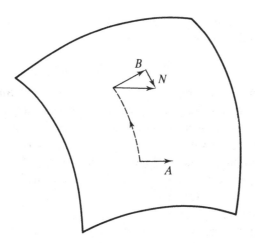

Fig. 3.1. Parallel displacement and the normal to a surface.

already made it convenient to operate in a manifold of four dimensions, three spatial and one temporal. (This had been a move, especially in the form given to it by Minkowski, calculated to produce all kinds of misplaced speculation amongst readers of popular science books. But it had no more significance than any other form of convenient description, and was simply based on the fact that the Lorentz transformation mixed together space and time variables.) So here again the four-dimensional case arose but naturally only the intrinsic properties would be available. The metric form now had, it turned out, ten components and the curvature was described, not by a single curvature number, but by a set of twenty. This set of numbers is called the 'Riemann–Christoffel tensor',[6] or RC tensor. I shall later need to say at some length what a tensor is and why it is important. For the present it is sufficient to regard the whole phrase as one technical term.

But how precisely was this to deal with gravitation? Starting with the special relativity situation in one dimension, for simplicity, there is a metrical form mentioned above:

$$ds^2 = dt^2 - dx^2/c^2.$$

Here dt, dx are the small changes in t, x coordinates and ds is the resultant change in proper time. An accelerated transformation such as

$$x \to x' = x - \tfrac{1}{2}gt^2,\, t \to t' = t,$$

generates a gravitational field g per unit mass. This field is uniform, that is, the same at all points and at all times. The transformation also produces a more complex and variable metrical form:

$$ds^2 = (1 - g^2 t'^2/c^2)dt'^2 - (2gt'/c^2)dt'dx' - dx'^2/c^2.$$

The metrical form *somehow* represents the gravitational field. But so far this is a bit of a hoax. It is only a uniform field, generated merely by changing our point of view (by being in a lift that suddenly accelerates upwards, say). It is a production of 'curvature' in the continuum of no more significance than, in the two-dimensional analogy, bending a piece of paper without crumpling. The geometry is basically unchanged and the intrinsic curvature, as measured by the RC tensor, remains zero. But this sets the stage for the final leap forward.

First, consider the previous argument in reverse. The field of uniform gravitation can be removed by an accelerated transformation. If the field is not uniform, but varying from point to point, one could still look at it in the immediate vicinity of some point. In this vicinity it is approximately uniform. So near any particular point the field may still be transformed away but this trick will not work at other nearby points. They will need their own slightly different transformation. For a varying field, the detailed mathematics confirms the commonsense guess that the price that has to be paid for this generalisation is that all these individual transformations cannot link up to give a single one which would transform away the whole field: the RC tensor is not zero. The upshot of the calculations is that a non-zero RC tensor corresponds to a non-uniform gravitational field.

Einstein's ideas were beginning to come together. The rotating disc had served its turn and could now be forgotten. The four notions muddled together by it were related like this: a gravitational field at some point *P* can be created or removed by an accelerated coordinate-transformation. The extent to which this transformation succeeds in doing the same thing at points near to *P* depends on the uniformity of the field or, equivalently, the vanishing of the RC tensor there. If the field is non-uniform, the geometry near *P* corresponds to an intrinsic curvature (analogous to the surface of a sphere in two dimensions) and is therefore non-Euclidean. Whether or not the field is uniform, the accelerated transformation produces new 'generalised' coordinates whose metrical meaning can be unpacked only by means of the metrical form. What were now needed were two further physical notions.

Firstly, this is meant to be a physical theory to determine the paths of particles moving in a gravitational field. What should be said about these particle paths? Because the geometry is four-dimensional, three spatial dimensions and one temporal, a physical 'path', that is, a succession of points occupied at successive instants of time, is represented by a single curve in the geometry. In the case of no gravitational field these curves are straight lines. One could infer that this will still be their *local* character when the local field has been transformed away. Because the local regions do not connect in the way that they would in the absence of gravitation, they will not produce from these locally straight lines an overall straight line. The local straightness, at *every* point will produce a path which is 'as straight as it can be'. Indeed, as one might guess, this criterion is just that the tangent direction to the curve is displaced along it by parallel displacement as explained above. These curves are what are called the 'geodesics' of the space.[7] In the analogy of the earth's surface, these straightest possible paths are the *great circle paths* of the navigators. We shall have a little more to say later about the geodesic nature of the paths.

The second, more serious one, is this. How is the metric form to be determined in a physical case? So far we have cheated by generating the complicated metric forms by transforming the coordinates. The problem in gravitation is posed quite differently. One is told that there are certain masses at certain positions and one has to determine their field, that is, their effect on a particle at any other point. Translating this into the Einstein language, one has to find how these masses determine the metrical form. The Gauss surface analogy is no longer any use here. One has to have recourse, instead, to the fact that the metric form is a description of the gravitational field. One hopes that it will be related to the masses in something like the same way that the Newtonian gravitational field is. Finding the correct 'field equations' took Einstein two more years. It is not necessary to go into detail about his struggle here. Basically, since the metric form has ten components, ten equations are needed and these equations must somehow involve the RC tensor. That has twenty components, so that equating it to zero would grossly overdetermine things (would, in fact, have only the solution of the uniform field). One needs only ten conditions and the struggle was in how to reduce the twenty components by combinations of them in some way to ten.[8]

Having done this, Einstein now had a theory describing the paths of particles in the field (determined by the field equations) generated by the other masses. That is, he had a gravitational theory. Physical applications became possible. The simplest case is the so-called *Kepler problem*, the determination of the orbits of the planets round a central gravitating sun.

Newton's gravitational theory says that the orbit of a single planet round a fixed sun is an ellipse with the sun at one focus. The observed motions are more complicated because the various planets feel the attraction of each other as well as the much larger attraction of the sun. The elliptical orbits rotate slowly as a result. The triumph of Newtonian mechanics in the eighteenth and nineteenth centuries was to explain these perturbations quantitatively to a high degree of accuracy. Only in the case of the planet Mercury, which is nearest to the sun, was the explanation less than accurate. The Newtonian prediction was of a rotation of 5557.18 seconds of arc per century as a result of the disturbances of the other planets. The observed rotation was of 5599.74 seconds. There was therefore a discrepancy of 42.56 seconds per century. When a solution of Einstein's field equations, corresponding to a central gravitating mass, was found by Schwarzschild in 1916, it predicted that, for a *single* planet in the situation of Mercury relative to the sun, the orbit would be an ellipse rotating by 43.03 seconds of arc per century. Since the observed discrepancy could be seen as the residual rotation after the effects of the other planets had been removed the near agreement of these two figures was regarded as a striking confirmation of the theory. The same solution of Schwarzschild also provided a calculation of the bending of light in a strong gravitational field. For a ray of light grazing the sun's disc, it predicted a bending through twice the angle predicted by Newtonian mechanics.

The eclipse expedition

So much for Einstein's paper. I have given a very partial sketch of the ideas behind it, stressing those which, explicitly stated or not, had the greatest effect on Eddington's thought. Now I turn to Eddington himself. He began by publishing a report (Eddington 1918) which introduced the subject to English-speaking physicists and astronomers. It also marked out Eddington as a leading expert on the new theory. This, together with his practical astronomical background, led to an incident which throws more light on his character. The details are to be found in *Eddington the Most Distinguished Astrophysicist of his Time* (Chandrasekhar 1983). An eclipse of the sun was to occur on 29 May 1919. This was an exceptionally good opportunity to test light-bending because on 29 May the sun is passing through an unusual patch of bright stars in the Hyades. In 1917 it was decided that the observations should be made, if the war allowed.

But there was more than interest in general relativity in this decision. After the Gallipoli fiasco had resulted in the death of Moseley, the British scientific establishment felt that something needed to be done to safeguard post-war science against the loss of its brightest practitioners. Conscription had been introduced. Eddington, at 34, was eligible but his Quaker background meant that he would refuse to serve. Conscientious objection was scarcely tolerated in England and so Larmor and Newall tried to have Eddington excused on the grounds of the country's long-term scientific interests. They succeeded in having the Home Office send a letter to Eddington which he had only to sign and return. But he was obstinate; after signing he added a postscript to say that he would not have served in any case, on grounds of conscience. 'Larmor and others were very much piqued.' The Home Office was legally obliged to act against known conscientious objectors. Only Dyson, by now the Astronomer-Royal, was able to save the day by having Eddington excused with the express stipulation of preparing for the eclipse expedition, if the war should be over.

There were two expeditions, sponsored by the Royal Society and by the government through the Greenwich Observatory, one to Sobral in Brazil and one to Principe, an island in the Gulf of Guinea. Eddington was in charge of the second. No instrument makers were available until after the Armistice and the expeditions had to sail in February. The rush and excitement certainly swept Eddington along, as they would anyone. They reached Principe on 23 April and settled in well. But 29 May saw a tropical rainstorm and cloud. The rain stopped at noon and the sun became visible only at 1.30 when the eclipse had already begun. But it was enough. By 3 June all the photographs were developed. The star images were disappointing because of cloud. But one of the more successful plates gave a fairly clear result and 'as the last lines of the calculation were reached, I knew that Einstein's theory had stood the test'. The confirmations of Einstein's prediction were reported by Dyson to a joint meeting of the Royal Society and the Royal Astronomical Society on 6 November 1919, with J.J. Thomson in the chair. There was tremendous publicity, the more so, as Rutherford remarked later, because the total destruction of the Edwardian social order was seen to have been followed by a new world in which it was possible to have the triumphal confirmation, by British astronomers in Brazil and West Africa, of an exotic prediction by a German physicist. Physics and astronomy were perhaps capable of transcending the old divisions that had caused such misery. In fact, later examinations of the eclipse observations have shown that the confidence with which they can be seen as confirming the theory is much less than it seemed in 1919, though later eclipses have provided more certain evidence.

None the less, Eddington told Chandrasekhar that, if he had been left to himself, he would not have planned the eclipse expeditions at all, so convinced was he of the correctness of general relativity. Whence came this certainty? In part, no doubt, from the first test of the theory, the motion of Mercury. Here there had already been a long series of astronomical observations. Their agreement with the new predictions was not in doubt so long as one was happy with the use of two different gravitational theories at once. One had to use Newton's theory to calculate and subtract the perturbations of the other planets and Einstein's to explain the residual rotation of Mercury's orbit. Most people simply regarded this as an approximately correct way of proceeding and Eddington was amongst them. It would just have been too complicated to calculate the perturbations by general relativity. But it would in any case have worried Eddington less than other people because of the sub-conscious belief in descriptive tolerance that I mentioned in the last chapter.

It was not Mercury alone that convinced Eddington and to find how he thought about the subject I shall examine in detail his book on the subject (Eddington 1923). Before I go on to this in the next chapter, it is worth noting his reaction to the so-called *third crucial test* of the theory. This was the prediction that light from atoms radiating in an intense gravitational field would be of a longer wavelength (lower frequency) compared with the same atoms in a weak field or none. This is the 'gravitational red shift'. This test is certainly an interesting one. In the first place, since the mere existence of the sharp spectral lines of light emitted by excited atoms requires an explanation by quantum mechanics, the test is a kind of check on whether general relativity and quantum mechanics have any common ground at all. This was a point to which Eddington was to return repeatedly. Secondly, the test depends critically on quantum mechanics in another way. It requires the assurance that the two atoms compared are exactly the same. The nature of this curious kind of agreement between quantum entities was not fully realised at the time. It is doubtful if it is fully understood even now. It is an exact agreement quite different in kind from the agreement between two cricket balls, where more careful measurement will always eventually find some small difference.

Einstein saw the third test as an important one for the historical reason that no such prediction had been made before the theory and so, of course, no observations had been made. It was a genuine prediction. Eddington was initially of the same opinion but, by the time he revised his *Report* for a second edition he had changed his mind:

But, if this test fails, the logical conclusion would seem to be that we know less about the conditions of atomic vibrations than we thought we did.

He had already moved to a position that, if there is a discrepancy in this no-man's-land between relativity and quantum mechanics, he is certain where the blame is to lie. So at the beginning of the 1920s Eddington was famous, with his early notions of descriptive tolerance and falsifiability intact, but with an addition. This was the affectionate certainty that general relativity, in the confirmation of which he played such a public role, was a sure foundation for the rest of science. Indeed, it had become more than a sure foundation. What empirical evidence could possibly falsify it, if the blame could always be placed elsewhere? Eddington was taking up a position very analogous to Kant, a philosopher with whom he had, he said, a great affinity. Kant took Newtonian gravitation as having apodeictic certainty. Most scientists in the twentieth century saw the overthrow of Newtonian gravitation by general relativity as simply confirming the refutation of the Kantian *a priori* synthetic which had begun with the discovery of non-Euclidean geometry in the nineteenth. Eddington, on the other hand, saw Kant as mistaken in his acceptance of Newtonian gravitation but not as a matter of principle. It was simply that Kant, in the eighteenth century took the wrong theory as certain. It was from such a point of view that he could now go on and write his text-book on the subject.

Notes

1. The equation of motion

$$\ddot{\mathbf{r}} = \mathbf{g}$$

becomes, if one transforms to a new frame of reference by

$$\mathbf{r} \to \mathbf{r}' = \mathbf{r} - \tfrac{1}{2}\mathbf{g}t^2 \ (t' = t),$$
$$\ddot{\mathbf{r}} = 0.$$

This is a familiar situation in text-books of Newtonian mechanics when rotating frames of reference are analysed (centrifugal and Coriolis forces).

2. It is easy to prove that passing a distance R from a gravitating point of mass M produces a deviation $2GM/Rc^2$, where G is the constant of gravitation and c the speed of light:

If $u = 1/r$ we have, using well-known formulae,

$$\frac{d^2u}{d\theta^2} + u = \frac{GM}{b^2c^2}, \text{ with } u = 0, \frac{du}{d\theta} = \frac{1}{b} \text{ when } \theta = \pi$$

giving the solution

$$u = (GM/b^2c^2)(1 + \cos\theta) + \sin\theta/b = (2/b)(\cos\theta/2)[(GM/bc^2)\cos\theta/2 + \sin\theta/2]$$

and $u = 0$ when $\cos\theta/2 = 0$ or $\tan\theta/2 = -GM/bc^2$. But the nearest approach is nearly at $\theta = \pi/2$, so that $u = 1/b$ and $R = b$, giving the result.

3. The name seems to result from a misguided way of expressing things. If one writes the equation of motion $\ddot{\mathbf{r}} = \mathbf{g}$ in the form $m\ddot{\mathbf{r}} = m\mathbf{g}$ and calls $m\mathbf{g}$ the gravitational force so as to compare with the other Newtonian forces, the equivalence in question is that between the inertial mass (the m on the left-hand side) and the gravitational mass (the m on the right-hand side).

4. Thus, $\mathbf{r} = \mathbf{r}(u^1, u^2)$ defines the surface, u^i $(i = 1,2)$ are coordinates, and $ds^2 = d\mathbf{r} \cdot d\mathbf{r} = g_{ij}du^i du^j$ where g_{ij} is written for the scalar product $\mathbf{r}_i \cdot \mathbf{r}_j$ ($\mathbf{r}_i$ being a partial derivative of $\mathbf{r}$ with respect to u^i) and a summation over $i, j = 1,2$ is implied whenever a literal suffix is repeated in a term.

5. Since $d\mathbf{r} = \mathbf{r}_i du^i$, the $\mathbf{r}_i$ serve as basis vectors and so their derivatives must be able to be written as:

$$\mathbf{r}_{i,j} = \Gamma_{ij}^p \mathbf{r}_p + L_{ij}\hat{\mathbf{n}},$$

where $\hat{\mathbf{n}}$ is the unit normal to the surface, Γ_{ij}^p are six ($\Gamma_{ij}^p = \Gamma_{ji}^p$) numbers called the coefficients of affine connexion and L_{ij} are three numbers called the coefficients of the first quadratic form, which determine the properties of the surface relative to the external space. Now

$$g_{ij;k} = (\mathbf{r}_i \cdot \mathbf{r}_j)_{,k} = \Gamma_{ik}^p g_{pj} + \Gamma_{jk}^p g_{pi},$$

which proves that

$$\tfrac{1}{2}(g_{ij,k} + g_{jk,i} - g_{ki,j}) = g_{jp}\Gamma_{ik}^p.$$

If g^{pq} is written for the inverse matrix to g_{jp} (assumed to exist) this equation can be solved in the form

$$\Gamma_{ik}^p = \tfrac{1}{2}g^{ps}(g_{is,k} + g_{sk,i} - g_{ki,s}),$$

so showing that the six Γ_{ik}^p are also intrinsic properties.

Any vector $\mathbf{A}$ in the surface, $\mathbf{A} = \mathbf{r}_i A^i$, has derivative in the $\mathbf{r}_j$ direction:

$$\mathbf{A}_j = \mathbf{r}_i A^i_{,j} + \Gamma_{ij}^p \mathbf{r}_p A^i + L_{ij}\hat{\mathbf{n}}A^i$$
$$= \mathbf{r}_p[A^p_{,j} + \Gamma_{ij}^p A^i] + L_{ij}\hat{\mathbf{n}}A^i.$$

The first part lies in the surface and is called the *covariant derivative*, $A^p_{;j}$. It is usual to extend the notion of covariant derivative by requiring:

(a) Leibniz's product rule,
(b) that the covariant derivative of a scalar function is its ordinary derivative,

(c) $g_{ij;k} = 0$.

It then easily follows that, if $B_i = g_{ij}B^j$, then

$$B_{i;j} = B_{i,j} - \Gamma^p_{ij}B_p,$$

and so for any entity behaving like a product of vectors extra Γ-terms are added and subtracted for each suffix.

The second part of the expression for A_j above is along the normal. Accordingly, if a vector is regarded as 'displaced' from a point x^a to a nearby point $x^a + dx^a$ in such a way that $dA^a = -\Gamma^a_{bc}A^b dx^c$, so that

$$A^a_{,c} + \Gamma^a_{bc}A^b = A^a_{;c} = 0,$$

this is called parallel displacement, or 'displacement without intrinsic change'. It would be better known as 'with as little change as possible'.

Now from

$$0 = \mathbf{r}_{i,jk} - \mathbf{r}_{i,kj} = (\Gamma^p_{ij,k} - \Gamma^p_{ik,j})\mathbf{r}_p + (\Gamma^p_{ij}\Gamma^q_{pk} - \Gamma^p_{ik}\Gamma^q_{pj})\mathbf{r}_q$$
$$+ L_{ij}\hat{\mathbf{n}}_{,k} - L_{ik}\hat{\mathbf{n}}_{,j},$$

it follows that

$$0 = R^q_{ijk}\mathbf{r}_q + L_{ij}\hat{\mathbf{n}}_{,k} - L_{ik}\hat{\mathbf{n}}_{,j},$$

disregarding further terms along the normal. The coefficients R^q_{ijk} are evidently intrinsic. To investigate the last terms, consider that $\mathbf{r}_i \cdot \hat{\mathbf{n}} = 0$, so that

$$L_{ij} + \mathbf{r}_i \cdot \hat{\mathbf{n}}_{,j} = 0.$$

Hence

$$0 = R^q_{ijk}g_{qs} - L_{ij}L_{sk} + L_{ik}L_{sj},$$

and defining $R_{sijk} = g_{sq}R^q_{ijk}$, it is clear that the only non-trivial equation is

$$R_{1212} = L^2_{12} - L_{11}L_{22}.$$

That is, the three measures of curvature, L_{ij}, are partly intrinsic but partly external; however, the determinant of the L_{ij} is an intrinsic measure of curvature, equally expressed by R_{sijk}, which later came to be called the RC tensor.

6. Given sufficient smoothness, the generalisation is easy. If the metric form is $g_{ij}du^i du^j$, then g_{ij} has $\frac{1}{2}n(n+1) = 10$ components. If a four-surface is embedded in N Euclidean dimensions, so that

$$\mathbf{r} = \mathbf{r}(u^1, u^2, u^3, u^4),$$

then the differential equations $g_{ij} = \mathbf{r}_i \cdot \mathbf{r}_j$ will be soluble if the N independent components of $\mathbf{r}$ are 10 in number. So $N = 10$ and there are consequently six independent normal directions $\mathbf{n}^p$ ($p = 1, \ldots 6$), which are most conveniently chosen as orthogonal to each other. There are then six first quadratic forms, L^p_{ij} and now $L^p_{ij} = -\mathbf{r}_i \cdot \mathbf{n}^p_{,j}$, so that

$$R_{sijk} = \sum_{p=1}^{6} (L^p_{ij}L^p_{sk} - L^p_{ik}L^p_{sj}).$$

It follows at once that $R_{sijk} = - R_{isjk} = R_{jksi}$, which reduces the number of possible components to $\frac{1}{2} \times 6 \times 7 = 21$ and it also follows that

$$R_{s(ijk)} = R_{sijk} + R_{sjki} + R_{skij} = 0$$

is one additional identity, reducing the possible number further to 20.

That there are, in general, no further algebraic identities is clear by considering a local coordinate transformation of the form $u \to v$ where

$$v^{i'} = b_i^{i'} u^i + \tfrac{1}{2} c_{ij}^{i'} u^i u^j + \tfrac{1}{6} d_{ijk}^{i'} u^i u^j u^k + \dots$$

near the origin. The $b_i^{i'}$, sixteen in number, allow the fixing of the g_{ij} at 0 to the special relativity form η_{ij}. The forty $c_{ij}^{i'}$ coefficients can then be arranged to ensure the vanishing of the forty first derivatives of the g_{ij} at 0. Next, the $4 \times (4 + 12 + 4) = 80$ $d_{ijk}^{i'}$ can be used to put eighty conditions on the hundred second derivatives. The remaining twenty degrees of freedom are those of R_{sijk} there.

7. The surface geometry analogue is helpful. In Note 5 above the derivative of a vector **A** splits into the covariant derivative, lying in the surface, and a part along the normal. Apply this to the tangent vector

$$\mathbf{t} = \frac{d\mathbf{r}}{ds} = \mathbf{r}_i \frac{du^i}{ds}$$

along the curve. The straightest curve is that in which, if dt/ds fails to vanish as it would without the constraint of the surface, at least it lies wholly along the normal (and so is 'not noticed' in the surface). This means that the covariant derivative vanishes in that direction,

$$\frac{d^2 u^i}{ds^2} + \Gamma_{jk}^i \frac{du^j}{ds} \frac{du^k}{ds} = 0$$

and the four-dimensional case is exactly the same. This is the differential equation for geodesics.

8. The Γ_{jk}^i in the geodesic equation involve first derivatives of the g_{ij} and evidently function somewhat like forces in Newtonian gravitation. The g_{ij} are therefore the analogues of the gravitational potential ϕ. The equation sought will therefore be generalisations of Laplace's equation $\nabla^2 \phi = 0$ or Poisson's equation $\nabla^2 \phi = 4\pi\rho$ of Newtonian gravitation as rewritten in the eighteenth century. There must be ten equations. R_{sijk} affords twenty combinations of second derivatives. However, $R_{ij} = g^{sk} R_{sijk}$ (a process called 'contraction' in the tensor calculus) combines these twenty into ten and one of Einstein's attempts at a field equation was $R_{ij} = 0$. Indeed, solving this except for a spherically symmetric singularity at the origin is sufficient to allow the predictions of the crucial tests. But even if correct, this is only the generalisation of Laplace's equation. When matter is present, bearing in mind the need for ten equations, one represents the matter by an energy tensor $T^{ij} = \rho (du^i/ds)(du^j/ds)$ (in the case of 'dust' with no internal structure) and Einstein tried the equations $R^{ij} = g^{is} g^{jt} R_{st} = T^{ij}$. This will not serve because the covariant divergence of T^{ij}, $T^{ij}_{\;\;;j}$ is identically zero but this is not

true for R^{ij}. Einstein's final form of field equations is best found as follows: the so-called Bianchi identities

$$R_{si(jk;m)} = R_{sijk;m} + R_{sikm;j} + R_{simj;k} = 0$$

are established by specialising the coordinate-system so that at one point O $g_{ij} = \eta_{ij}, \Gamma^i_{jk} = 0$ and so covariant derivatives are ordinary derivatives and

$$R_{sijk;m} = \eta_{sp}(\Gamma^p_{ij,km} - \Gamma^p_{ik,jm}).$$

Now Γ^p_{ij} is defined in terms of the first derivatives of the g_{ij} and the substitution of the values into the right-hand side at O gives an expression in terms of third derivatives which obviously vanishes as a result of the cyclic permutation. Multiplying the identities by g^{sk} gives

$$R_{ij;m} - R_{im;j} + R^k_{imj;k} = 0.$$

Multiply this by g^{ij} and define $g^{ij}R_{ij} = R$:

$$R_{;m} - R^j_{m;j} - R^k_{m;k} = 0,$$

that is,

$$R^p_{q;p} = \tfrac{1}{2}R_{;q} = \tfrac{1}{2}\delta^p_q R_{;p},$$

so that $[R^p_q - \tfrac{1}{2}\delta^p_q R]_{;p} = 0$. One therefore takes $R^{pq} - \tfrac{1}{2}g^{pq}R$ as a multiple (to allow for choice of units, since T^{pq} is a physically defined quantity) of T^{pq} and this is Einstein's law of gravitation in 1916.

4

Consequences of general relativity

As well as changes in his life, general relativity brought deep intellectual consequences for Eddington. In this chapter I make a significant step towards understanding the Eddington mystery by tracing the results of his immersion in the elegant theoretical construct of general relativity and his successful confirmation of its predictions. This chapter and the next are two complementary philosophical discussions. The first of these sets out eight features of Eddington's later thought, some acknowledged by him, some only implicit. The second deals at some length with his one major error about general relativity, his assumption of the universal character of tensors. This second argument has a slightly technical aspect, which I have reduced as much as possible. What remains is vital in understanding why Eddington made a mistake the revelation of which in 1928 had such a profound psychological effect on him.

Matter as a construction

With the eclipse expedition behind him Eddington began to publish the many thoughts that his experience in general relativity had produced. It was an amazing explosion of intellectual effort. Some of these publications were explanations of the theory in a more or less technical manner but he also began to write on the philosophical consequences he saw flowing from it. It is dangerous to take Eddington's philosophical writings at face value. When he isolates certain ideas as the important ones for understanding physics it is often because he takes for granted others which are, in fact, more idiosyncratic and more in need of exposition. Dingle, in a generally sympathetic obituary notice of Eddington (Dingle 1945), remarked that

In most matters the groove of his thinking was cut so deep that only with difficulty, if at all, could the thoughts of others enter it . . . I would venture the opinion that his failure to convince others was due to that mental peculiarity to which I have already alluded, which prevented him from understanding the way of thinking common to most of his colleagues.

If that is true, and I think it is, it illustrates the size of the task before us in sorting out what made Eddington write his last books.

From the point of view of his development, however, it is useful to look in some detail at a lengthy paper in the philosophical journal *Mind* (Eddington 1920b). The ideas in this paper were not all to be retained indefinitely by Eddington but the paper is important in showing his immediate reaction to relativity. He contrasts theoretical structures with the world of perception. Then he makes a disarming disclaimer:

There are in fact a number of possible sites for a bridge between the analytical theory and the phenomena of perception. As has been said, the physicist commonly makes the connexion through things that are measured experimentally. Another alternative is to carry on the analytical development of the external world to the point at which it meets mind in the nerve-centres of the brain. In this paper I have taken the middle course of making the connexion through the everyday world which we see and feel around us.

One might well ask, if such a middle ground is available, why is it not occupied by almost everyone in the two other camps?

Eddington then goes on to describe his own approach to the beginnings of (special) relativity (use of slow transport of clocks and a general argument that relations between events, i.e. proper-time lapse, are the basis). Next comes the main argument of the paper to the effect that matter is not a substance with real predicates discovered by experiment. Rather it is an *a priori* mathematical construction from relations between measurables. It is not too difficult to see how he comes to this point of view but less easy to understand why he regards it as characteristic of relativity.

The notion really refers to the formulation of the field equations in the theory. Without going into unnecessary detail about these, one can express the argument equally in a version of Newtonian theory, as that was itself reexpressed in the eighteenth century. There the gravitational field at a point, $\mathbf{g}$ say, is expressed in terms of the variation of a numerical quantity ϕ

depending only on position, called the gravitational potential, by the equation $\mathbf{g} = -\nabla\phi$. This is a shorthand way of saying that the three components of the field g along the three coordinate directions are the three partial derivatives $-\partial\phi/\partial x$, $-\partial\phi/\partial y$, $-\partial\phi/\partial z$. Laplace showed that in order that the potential be a description of Newton's inverse square law a certain condition, C (in fact, though the details do not matter, Laplace's equation $\nabla^2\phi = \partial^2\phi/\partial x^2 + \partial^2\phi/\partial y^2 + \partial^2\phi/\partial z^2 = 0$) is placed on the variations of the potential from place to place. Later Poisson showed how to modify this condition when matter is present. He did this by imagining a very small cavity to have been cut out at the point of the matter in which we are interested. Inside this cavity there is then no matter, so that the condition C applies. But one has also to attend to the boundary of the cavity and to the conditions on the potential across the surface of it. When one does this, one finds that the result when the cavity is small enough is the same as if the cavity were not there but the condition C^* (in fact, Poisson's equation $\nabla^2\phi = 4\pi Gm$, where G is the constant of gravitation and m is the local density of matter) replaces the condition C. This shows how, by a trick as it were, the two cases of free and filled space can be considered together. As it is usually put, at a point where there is no matter, $m = 0$ and C^* reduces to C.

The same kind of argument comes up in general relativity; the obvious difference is that it is now the ten components of the metric form that have to be determined, so that C now comprises a ten-fold condition, as does C^*. In the latter case, then, m must be replaced by something describing the matter with ten components, and this is known as the energy-momentum tensor. It is exactly at this point that Eddington reverses the argument and takes the ten components of the left-hand side of C^*, now called the Einstein tensor, as logically prior. If it is non-zero, he argues, then it *defines* matter by giving its energy and momentum. If it happens to be zero, there is no matter present. So C is not a law of nature after all, as it was in Newtonian mechanics, but simply a statement that matter is absent. This argument is what gives Eddington the basis for his view that matter is not a substance discovered by experiment but a construction.

Why does Eddington take this 'reversed interpretation' as characteristic of relativity? Evidently not for its *logical* structure, which simply replaces the identical structure in Newtonian theory. In general relativity there are three further features that influenced him. In the first place, he had been able to rewrite the beginnings of Einstein's theory in such a way as to present it as a more or less inevitable consequence of an analysis of structure and relations. This would appear later in his text-book (Eddington 1923). From these beginnings the progress to the Einstein tensor was, as he says with pardonable optimism, 'in the main, deductive'. The qualification 'in the main' covers a multitude of sins but the phrase is true enough to make the reversed interpretation more natural here than in the Newtonian case. Secondly, the Einstein tensor satisfies a *condition of continuity* which is somewhat like the condition[1] satisfied by the mass and momentum of a fluid in order that it represents a flow (that is, no cavities or sudden sources of matter). It is not exactly the same condition since the Einstein tensor has ten components rather than four but there is a close analogy. Now the consequence of this condition of continuity on the energy tensor is to specify the way in which the matter moves. In later years this fact came to be known as the determination of the equations of motion by the field equations. It was later fully investigated even for the extreme case of the motion of two point singularities by Einstein and his collaborators (Einstein, Infeld and Hoffman, 1938). Eddington does not put it like this but he was certainly aware of the simpler case when the matter has the form of 'dust', that is, non-interacting small particles. This determination of the equations of motion points to the fundamental role of the Einstein tensor, and because this is missing altogether in Newtonian gravitation, it provides a second justification for Eddington's view that his reversed interpretation is a consequence of relativity.

However, the third aspect of relativity that motivates Eddington's argument is the crucial one for him. He remarks, after an explanation of his somewhat conventionalist foundation for special relativity starting from the notion of interval,

But in a strict analytical development the introduction of scales and clocks before the introduction of matter is – to say the least of it – an inconvenient proceeding. Thus in our development [the Einstein tensor] is not merely of unknown nature but unmeasurable.

Eddington is sure that his foundations for special relativity are the best ones. This then requires him to bring in matter earlier than he wishes and so this forces him, in turn, into an idealistic stance. It is all in line with Kant's view of a real external world, of which our experience gives us no knowledge of things-in-themselves. Eddington argues that physical reality 'in itself' is provided in the theory by an amorphous space-time continuum which is inaccessible to direct measurement. He sees no ontological consistency in this reality over and above what is provided by the mind conceiving it:

The intervention of mind in the laws of nature is, I believe, more far-reaching than is usually supposed by physicists. I am almost inclined to attribute the whole responsibility for the laws of mechanics and gravitation to the mind, and deny the external world any share in them.

Such a forthright idealism could hardly survive; it must be seen as an over-enthusiastic reaction to the many strange and exciting features of the theory. But it left him with a view about laws of nature which was to stay with him:

Are there then no genuine laws of the external World? Is the universe built from elements which are purely chaotic?

It can scarcely be doubted that our answer must be negative. There *are* laws in the external World, and of these one of the most important (perhaps the only law) is a law of atomicity.

Later in the year Eddington took part in a symposium which is also reported in *Mind* (Eddington, 1920c) in which he tells an amusing story comparing the scientist to an 'aged college bursar' cut off from life except for his attention to the college bills. 'He vaguely pictured an objective reality at the back of it all – some sort of parallel to the real college'. Then he notices the double-entry nature of the accounts. 'It is a perfect law of nature with no exceptions. Credit must be called plus and debit minus; so we have the law of conservation of £.s.d.'

The *Mathematical Theory of Relativity*

These then are the philosophical ideas that Eddington put forward in 1920, but his real philosophical position had crystallised considerably by the time he came to write *Mathematical Theory of Relativity* (Eddington 1923) which I shall refer to as hereafter as MTR, and it is to the ideas of that book that I now turn. It was hailed at the time as a definitive text-book on a new subject, and it is certainly that. Until recently it held its own as one of the clearest expositions and it is only because of the many new developments in general relativity since 1945 that it has been replaced by swifter treatment of the fundamentals over which it lingers. The technical details of the book do not concern me much. About three-quarters of it is an exposition of the mathematics of Einstein's 1916 paper on more or less the same lines as Einstein's, though the arguments behind the mathematics are sometimes different. There is a difference of emphasis; Eddington tends to put a greater emphasis on the concept of curvature than Einstein does. This chimes better with modern thought than Einstein's approach. The alternative path favoured by Einstein is to interpret the theory physically in terms of the different elements in the equation of motion of a particle.[2] But locally these elements can be reduced to zero by a transformation of the coordinate-system to a freely-falling one. They cannot then be an intrinsic description of the gravitational field. In fact, they simply describe the coordinate-system adopted, although what coordinate-systems can be adopted does depend on the gravitational field, so that the field is described indirectly. The situation is analogous to the choice of coordinate-systems to describe the behaviour of people trying (in the old fair-ground set-up) to stay on a horizontal and fairly smooth circular rotating turntable. The 'natural' reference-frame for the participants is one fixed to the disc and so, in it, they experience a force pulling them outwards until it overcomes the friction and they fly off. But the onlooker, on the ground, more naturally takes a reference frame fixed to the ground and he sees the effect of the rotation of the disc as producing a tendency of the participants to cease the inward acceleration towards the centre produced by the rotation of the disc and, once the friction is overcome, to move straight ahead. There is no radial force for him. On the other hand, a description of gravitation in terms of the curvature cannot be removed by a coordinate transformation.

The remaining quarter of the book deals with two attempts to incorporate electromagnetism into the same kind of framework that relativity has provided for gravitation or, as one would now say, two unified field theories. It is surprising that such a thing should have been attempted. After all, the whole structure of general relativity flows from the distinctive properties of gravitation, properties not shared in any way by electromagnetism. I suppose that, in the historical context, it is the obverse of the development of general from special relativity. Once that had been achieved, it seemed as if electromagnetism, which had played such an important part in special relativity, had retreated into the background. Weyl, Eddington and, later, others seem to have felt that things had not been done quite correctly, that the balance needed redressing. The first theory described in the book is that of Weyl and the second is Eddington's generalisation of it. Both theories proceed by generalising the notion of parallel displacement but neither generalises this as far as it can be taken. It is for this technical reason that neither of the theories was successful[3] and Eddington later came to the view that unified field theories could not succeed. The complete generalisation was not carried out for a quarter of a century but even when it was, the resultant theory appears to have no clear physical significance.[4]

Leaving these technical aspects on one side, then, what are the ideas that inform Eddington's thinking in MTR? The answer to this is complicated, because it comes from a mixture of explicitly avowed philosophical views, some foreshadowed in *Mind* (Eddington 1920b), some different, with a greater variety of implicit views which are never avowed. I will try to clarify the situation by listing eight such views to be found in MTR, together with detailed references. These ideas are not all independent but, because Eddington held most of them only implicitly, it is pointless for me to discuss their logical connexions, coherence and so on. By listing them here I do not mean to imply any opinion about their validity or usefulness. It is sufficient for the present discussion that Eddington held them. The references are to the pages of the second edition of MTR of 1924, as this is readily available in its various reprints. It does not differ materially from the first edition.

I have listed the views as they arise but a word is necessary about their different status. The views put forward explicitly by Eddington represent his notion of how physicists actually work, though they may well not know this and may disagree with him. Once he felt he had isolated such an idea, his intention was to use it openly. The more important views are not put forward explicitly. They arise from a careful reading of Eddington's descriptions of his technical procedure. That is to say, they come from Eddington's preconceptions of the scientific enterprise. It will transpire that some of these preconceptions are of a surprising nature. Several have not been noticed before. The reader may wonder what method of procedure has enabled me to be so certain when others have failed to notice them. The answer is simple; the ideas are clearly implicit in MTR to anyone who returns to it after studying (Eddington 1936, 1946). I am told that many years ago G.C. McVittie, who had been a student of Eddington, insisted in conversation on the strangeness of many of the ideas of MTR. Perhaps he stimulated my thinking, for he was my supervisor, but I do not remember it.

1   *Selective subjectivism*

I have already drawn attention to one of the explicit views, that of selective subjectivism, a view later to be encapsulated by Eddington (1939) in his story of the ichthyologist who catches all his specimens by a particular net and then writes a paper 'On the minimum size of fishes'. He also captures it in another story of an aprocryphal paper of Procrustes: 'On the uniform length of travellers'. The story of the old college bursar is also relevant.

2   *Descriptive tolerance*

This is one of two implicit ideas to which I drew attention before. One might reasonably associate this with the name of Wittgenstein, so long as it is understood that no direct influence is intended. Wittgenstein did not return to Cambridge until 1929 and the idea of a language-game appears in his lectures, written up in the Blue Book of 1933. I know of no influence of Wittgenstein's teaching on Eddington. We are in any case talking of Eddington's thought in the early 1920s, though it becomes more explicit by 1936. Descriptive tolerance has the form of describing a body of scientists in one discipline as playing a language-game. They do not *have* to play that particular language-game and those in a different but related discipline may well be playing another, although sometimes describing the same phenomena.

It is fair to see these first two views of Eddington as leading him towards a naïve relativism of rather an extreme kind. But this must be taken alongside the distinction he makes repeatedly in MTR (see 6 below) between quantum physics and the rest.

3   *Falsifiability*

The other implicit idea to which I have already drawn attention might similarly be associated with the name of a philosopher, in this case Popper. Here again no suggestion of influence is implied. It is simply that Eddington acts with a rather stronger dose of what he regards as scientific common-sense than most of his colleagues. In at least two places in (Eddington 1926) he uses the criterion without explicit reference. One such place is his refusal to drop the ideal gas law in the descending series in stellar evolution, and the other is in his choice of the only two observable parameters as controls in his stellar theory.

4   *Operationalism*

Here again I associate Eddington's idea with a philosopher – Kant. Eddington has himself said that Kant was a philosopher with whom he felt a sympathy. Consider the following passage:

That all our *physical* knowledge begins with *experiment* there can be no doubt. For how is it possible that the faculty of *theorising* should be awakened into exercise otherwise than by means of objects which affect our senses . . .. In respect of time, therefore, no knowledge of *physics* is antecedent to *experiment*, but begins with it.

But, though all our knowledge begins with *experiment* it by no means follows, that all arises out of *experiment*. For, on the contrary, it is quite possible that our empirical knowledge is a compound of that which we receive through impressions, and that which the faculty of *theorising* supplies from itself (sensuous impressions giving merely the occasion), an addition which we cannot distinguish from the original element given by sense, till long practice has made us attentive to, and skilful in separating it.

I have given no reference for this passage, for it is a made-up affair got by taking some of the 'Introduction' of Kant's *Critique of Pure Reason* (in Meikeljohn's translation) and changing a few words which are italicised. My reason for this is that the amended passage captures Eddington's operationalism so perfectly. He himself sets it out very clearly in the last paragraph of MTR (p. 240):

The physicist who explores nature conducts experiments. He handles material structures, sends rays of light from point to point, marks coincidences, and performs mathematical operations on the numbers which he obtains. His result is a physical quantity, which, he believes, stands for something in the condition of the world. In a sense this is true, for whatever is actually occurring in the outside world is only accessible to our knowledge in so far as it helps to determine the results of these experimental operations. But we must not suppose that a law obeyed by the physical quantity necessarily has its seat in the world-condition which that quantity 'stands for'; its origin may be disclosed by unravelling the series of operations of which the physical quantity is the result. Results of measurement are the subject-matter of physics; and the moral of the theory of relativity is that we can only comprehend what the physical quantities *stand for* if we first comprehend what they *are*.

It will be noticed that the word *are* is being used in a special way here. Eddington qualifies his view a little more fully on the first page of MTR:

To find out any physical quantity we perform certain practical operations followed by calculations; the operations are called experiments or observations according as the conditions are more or less closely under our control. The physical quantity so discovered is primarily the result of the operations and calculations; it is, so to speak, a *manufactured article* – manufactured by our operations. But the physicist is not generally content to believe that the quantity he arrives at is something whose nature is inseparable from the kind of operations which led to it; he had an idea that if he could become a god contemplating the external world, he would see his manufactured physical quantity forming a distinct feature of the picture.

and somewhat more on p. 3:

The study of physical quantities, although they are the results of our own operations (actual or potential) gives us some kind of knowledge of the world-conditions, since the same operations will give different results in different world conditions. It seems that this indirect knowledge is all that we can ever attain, and that it is only through its influence on such operations that we can represent to ourselves a 'condition of the world'.

Eddington's reasons for emphasising this view are two-fold. In the first place he is concerned to disarm any critics of the special relativistic modification of the notion of time which renders it relative to an observer. Such a modification is more naturally accommodated by the operational point of view in which the definition simply arises from the observer's rules for the determination of time. Secondly, he sees operationalism as a powerful argument against those who baulk at learning the rather complex mathematical techniques, the 'tensor calculus', needed for the general theory: 'our knowledge of conditions in the external world, as it comes to us through observation and experiment, is precisely of the kind which can be expressed by a tensor and not otherwise.' This last view is something to which we must return in the next chapter, where I explain at some length what a tensor is. It happens to be false and it is this fact that enabled Dirac to express the wave equation for the electron in 1928 in a form which held for all inertial observers and yet was not of tensor form. Eddington's conviction of the truth of the false statement, a belief he held in common with the general community of physicists, led to his profound shock in 1928.

I have now described Eddington's notion of operationalism. I must make it clear that this view is substantially different from a more extreme one which is known by the same name. This more extreme view, usually associated with Bridgman, holds, in its original form, that physical quantities are wholly defined by the operations of their measurement and that there is no more 'reality' than that. (Bridgman's later views were more subtle.) Eddington readily accepted the idea of an operationalist definition of a measurement but he strongly felt the common-sense view that there were 'conditions of the world' that were being measured. Indeed Dingle in his obituary (Dingle 1945) regarded it as one of Eddington's self-imposed difficulties that he strove to stake a middle course between two well-trodden paths. Dingle sees the non-operationalist one of these as the old pre-relativistic view that physics is simply discussing the objective features of the world. The other view is that of the operationalists, in the Bridgman sense. Dingle says:

The inscrutable 'conditions of the world' hung like the Old Man of the Sea round the neck of his thought, contributing nothing and serving only to retard its progress and obscure results which, expressed simply and directly in terms of the essential measurements alone, might have commanded understanding and acceptance.

I think that Dingle is probably mistaken in the judgement he makes on this aspect of Eddington's thought. At any rate, it was of the greatest importance to Eddington to stand out against the austere operationalism described by Dingle. Otherwise Eddington's attitude to the third crucial test of general relativity becomes untenable. If the test should fail, there must be external conditions of the world (that is, quantum mechanics) to take the strain. Eddington's view becomes very clear on p. 5 of MTR, where some modifications of the notion, which play an important part in his later work, are set out:

Any operation of measurement involves a comparison between a measuring appliance and the thing measured. Both play an equal part in the comparison and are theoretically, and indeed often practically, interchangeable; for example, the result of an observation with the meridian circle gives the right ascension of the star or the error of the clock indifferently, and we can regard either the clock or the star as the instrument or the object of measurement. Remembering that physical quantities are results of comparisons of this kind, it is clear that they cannot be considered to belong solely to one partner in the comparison.

Taking this a little farther on the same page:

We have seen that the world-condition or object which is surveyed can only be apprehended in our knowledge as the sum total of all the measurements in which it can be concerned; any *intrinsic* property of the object must appear as a uniformity or law in these measures. When one partner in the comparison is fixed and the other partner varied widely, whatever is common to all the measurements may be ascribed exclusively to the first partner and regarded as an intrinsic property of it. Let us apply this to the converse comparison; that is to say, keep the measuring-appliance constant or standardised, and vary as widely as possible the objects measured – or, in simpler terms, make a particular kind of measurement in all parts of the field. Intrinsic properties of the measuring appliance should appear as uniformities or laws in these measures. We are familiar with several such uniformities; but we have not generally recognised them as properties of the measuring-appliances. We have called them *laws of nature*!

It will not do to follow Dingle in lamenting the complexities of Eddington's operationalism. The complex idea remains part of his thinking and motivates his later work.

5a   Theory-languages: Case-histories

The fifth view was not openly acknowledged by Eddington and was only brought out into the open by E. W. Bastin in the 1950s (Bastin and Kilmister 1954). I follow him in calling it the notion of a theory-language. There is some common ground with the discussion above of descriptive tolerance; such tolerance is a necessary condition for the existence of theory-languages, but this fifth view is a more subtle limitation on theoretical expression. I begin with some detailed background. On p. 2 of MTR Eddington says:

The parallax of a star is found by a well-known series of operations and calculations. The distance across the room is found by operations with a tape-measure. Both parallax and distance are quantities manufactured by our operations; but for some reason we do not expect parallax to appear as a distinct element in the true picture of nature in the same way that distance does. Or again, instead of cutting short the astronomical calculations when we reach the parallax, we might go on to take the cube of the result, and so obtain another manufactured quantity, a 'cubic parallax'. For some obscure reason we expect to see distance appearing plainly as a gulf in the true world-picture; parallax does not appear directly, though it can be exhibited as an angle by a comparatively simple construction; and cubic parallax is not in the picture at all.

Another aspect is shown on pp. 6–7 where Eddington is involved in the possible limitation of the operational measurement of mass at very short distances: he would be puzzled, he says, to give the exact operational means of determining a very small length (of 10^{-13} cm) but he intends to ignore the difficulty. To the possible criticism that this is returning to the bad old ways he replies:

By all means explore this criticism if you regard it as a promising field of inquiry. I here assume that you will probably find me a justification for my 10^{-13} cm.; but you may find that there is an insurmountable ambiguity in defining it. In the latter event you may be on the track of something which will give a new insight into the fundamental nature of the world.

A third case-history is to be found on pp. 166–7 of MTR in a discussion of the so-called Einstein universe in relativistic cosmology. It is useful to look into the background a little here, as the Einstein universe plays an important role in Eddington's later thought. Within a couple of years of the discovery of the field equations, both Einstein and de Sitter began to look for solutions which would correspond to (a smoothed-out simplification of) the whole universe. Such a solution would be one with matter present. In the discussion above about the field equations in matter, the energy-momentum tensor (the ten components of which were written symbolically as m) on the right-hand side of the condition C^* describes the nature of the matter. Einstein considered it very important to have a static non-empty solution. This was before the discovery of the expansion of the universe, so that a static world was regarded as more important than it is in modern cosmology. In this static non-empty case m would have a special form. The assumption is made that this corresponds to a uniform density of matter (because the model smoothes out the matter and, since all points of the universe are regarded as equivalent, the density is uniform). The form of *m* must also correspond to zero velocity. The original equations, *C**, do not have such a static, non-empty solution. Einstein noticed, however, that the Einstein tensor, which I denote symbolically by *G* (so that the condition *C** is *G = m*), could be modified by adding a term, called 'the cosmical term' proportional to the metric coefficients, *G + kg*, where *k* is a constant called the cosmical constant. The important point which justifies this move is that the new, modified Einstein tensor still enjoys the property of continuity mentioned above. So the condition in matter is no longer *C** but, say, *C*+*, which denotes *G + kg = m*. One consequence of this is that the empty space equations, *C*, will also be modified to *C+*, that is *G + kg = 0*. If the accuracy of the predictions of the advance of perihelion of Mercury is not to be lost, *k* must be sufficiently small.

The conditions *C*+*, together with the non-emptiness, turn out to be just enough to determine a unique solution, called the Einstein universe. One consequence of this solution is a relation between the total mass *M* of the universe, a quantity which represents a length (proportional to its total size) in it, *R* and the cosmical constant *k*:

$$M = \tfrac{1}{2}\pi R = \tfrac{1}{2}\pi/\sqrt{k}.$$

I now return to Eddington's argument. This arises when he is discussing, in MTR, a view of Einstein, that this relation, though proved for the Einstein solution, is not specific to this actual model, but would hold for any universe. Such a view would not be entertained now for a moment, but that is unimportant for our purposes. Eddington first asks whether there could be some physical mechanism which would adjust *k* when *M* increased, so as to keep the equation true; this seems to him impossible:

But the suggestion is perhaps more plausible if we look at the inverse relation, *viz M* as a function of *L*. If we can imagine the gradual destruction of matter in the world (e.g. by coalescence of positive and negative electrons), we see by (71.1) that the radius of space gradually contracts; but it is not clear what is the fixed standard of length by which *R* is supposed to be measured. The natural standard of length in a theoretical discussion is the radius *R* itself. Choosing it as unit, we have $M = \tfrac{1}{2}\pi$, whatever the number of elementary particles in the world. Thus with this unit the mass of a particle must be inversely proportional to the number of particles.

Later he goes on to say:

In favour of Einstein's hypothesis is the fact that among the constants of nature there is one which is a very large number; this is typified by the ratio of the radius of an electron to its gravitational mass $= 3 \cdot 10^{42}$. It is difficult to account for the occurrence of a pure number (of order greatly different from unity) in the scheme of things; but this difficulty would be removed if we could connect it with the number of particles in the world – a number presumably decided by pure accident. There is an attractiveness in the idea that the total number of the particles may play a part in determining the constants of the laws of nature; we can more readily admit that the laws of the actual world are specialised by the accidental circumstance of a particular number of particles occurring in it, than that they are specialised by the same number occurring as a mysterious ratio in the fine-grained structure of the continuum.

5b   *Theory-languages: Applications*

I now try to trace the notions common to these examples. In the first case, the 'obscure reason' is that the scientists are operating in a peculiarly restricted language-game. Such a restricted situation is what is meant by a theory-language. Some mathematical operations are allowed (such as those needed in converting between parallax and distance), others are forbidden (cubing the parallax). The same sort of restriction is invoked in the last of the passages. It refers to a 'very large pure number'. This number, quoted as 3×10^{42} appears as a ratio of two quantities associated with an electron. Eddington's argument to connect it with the number of electrons can only have any logical validity in a theory-language in which the ratio in question cannot, for example, have its logarithm taken, since then it would not be a particularly large number, but simply about 42.5.

It is interesting to note that such an orthodox physicist as P. A. M. Dirac tacitly uses the same concept (Dirac 1938). Here, following up his earlier idea in *Nature* (Dirac, 1937), he is concerned to explain an 'empirical fact' about certain constants, independent of the system of units, that arise mainly in cosmology. For example, the ratio of the electric force e^2/r^2 to the gravitational force Mm/r^2 (M the proton mass, m the electron mass) is a ratio of two forces and so is the same whatever units are employed to measure force. The number is of the order of 10^{40}. There are also other numbers of roughly this size. Again, the ratio of the masses of the proton to that of the electron, M/m is roughly 1836. Dirac considered all the 'dimensionless constants' of this kind, allowed himself the latitude of replacing any one less than unity by its reciprocal, and observed that the resulting collection of numbers were in three groups. One group was of magnitude roughly unity (up to a few factors of 10; an example is M/m), one of magnitude roughly 10^{40} and one of magnitude around 10^{80}. The conclusion that Dirac draws from this ('In some ways . . . a counsel of despair' according to a leading cosmologist (Bondi, 1960)) does not concern us; he simply assumes that the elements in the second group contain the age of the universe (in certain natural units) as a factor and those in the third group its square. What is of interest is the implied assumption, not stated by Dirac, that arbitrary mathematical transformations are not to be permitted. For if we happen to redefine symbols in some way so that $e/\sqrt{(Mm)}$ enters instead of e^2/Mm, the new constant will have the value 10^{20}, and will not fall into any of the groups.

Returning to Eddington's notions, the idea of two theory-languages, in one of which it is taken for granted that lengths of 10^{-13} cm can enter, whatever the difficulties over operational definition, whilst in the other the measurement of such lengths is investigated, but with the two theory-languages equally valid, makes sense of Eddington's contention. The point here is that the theory-languages have different degrees of complexity. To each such theory-language corresponds a certain body of experiment. It would be an invalid procedure to use experiments from one theory-language to criticise results in another, simpler, one.

Lastly, the theory-language concept illuminates Eddington's curious argument about the absence of a standard of length by which the contraction of R can be measured. This argument seems bizarre because it comes at the end of a long chain of argument in which it might well have been assumed in the earlier stages that appropriate standards of mass, length and time were lurking in the background, just as in any other physical investigation. Of course, it is invalid to bring in the theory-language concept at the end; it is necessary for it to motivate the whole investigation. The investigation might then be very difficult to carry out since, except in certain very special cases, the exact rules of a theory-language are difficult to specify. This is perhaps the first appearance of a characteristic Eddingtonian difficulty. New notions are sometimes stated, but more often implied, at a late stage of an investigation. It is not clear whether or not the earlier stages of the investigation depend critically on the validity of conventional notions, inconsistent with the new ones. What is clear from these examples, however, is that the notion of a theory-language was present in Eddington's thought already by 1923. That it is nowhere stated may be because he was unaware of it or because he took it for granted that everyone held the same view. In either case it must be taken into account in considering his later work.

6   *Physics as structure*

The notion of structure rather than substance as the subject matter of physics is said by Eddington to have come from his reading of Russell, who argues for the corresponding statement about mathematics. By 'structure' Eddington means a complex of relations and their terms. One aspect of this view has already surfaced in my description of the *Mind* paper. The notion of matter as not a substance but an *a priori* mathematical construction from relations is exactly in line with this view. The more general view is explicitly stated on p. 41 of MTR:

The investigation of the external world in physics is a quest for *structure* rather than *substance*. A structure can best be represented as a complex of relations and relata; and in conformity with this we endeavour to reduce the phenomena to their expressions in terms of the relations which we call intervals and the relata which we call *events*.

This is used to substantiate Eddington's preferred beginning for relativity theory in terms of the proper time, viewed as a numerical relation between events. The emphasis on structure is also involved in Eddington's Kantian philosophy that I mentioned above. That it remained with him can be seen in a review of *Philosophy of Physical Science* (Eddington 1939) in *Mind* (Braithwaite 1940):

Eddington's language and emphasis on epistemology throughout the book make it clear that his contention is that the fundamental laws of physics are *a priori* in Kant's sense, and that the reason for their apriority is essentially Kant's reason for the apriority of his synthetic *a priori* judgements – that they are presupposed in the way we make judgements.

But Braithwaite goes on to point out a most important qualification on Eddington's view. It is only the *fundamental* laws and so on, not 'the vast amount of special information about the particular objects surrounding us' which is merely structure. There is undoubtedly an element of tautology here. No independent criterion is given to enable the reader to judge what is fundamental and therefore structural. But this is not important; the point being made is that there is some considerable body of physics that falls under both headings. It would be highly misleading to interpret Eddington here as regarding physics as making an important contribution to the understanding of the everyday world by showing that it is in some way devoid of substance. His view is very much more that this fundamental structural knowledge which physics provides, fascinating toy though it is to explore, is only something to be cleared out of the way to allow a better understanding of God's handiwork. In this he stands in sharp contrast to his great predecessor, Newton, who saw natural philosophy as illuminating the Divine.

There are limitations to this structural knowledge even within physics. On p. 153 of MTR:

The possibility of the existence of an electron in space is a remarkable phenomenon which we do not yet understand. The details of its structure must be determined by some unknown set of equations, which apparently admit of only two discrete solutions, the one giving a negative electron and the other a positive electron or proton.

This shows again Eddington's recurring thought that quantum theory may contain the only non-conventional law of nature.

The clearest expression of this sixth view of Eddington is found in his summing-up of his unified field theory on p. 224 of MTR, which would have had Kant's approval:

The fundamental basis of all things must presumably have *structure* and *substance*. We cannot describe substance; we can only give a name to it. Any attempt to do more than give a name leads at once to an attribution of structure. But structure can be described to some extent; and when reduced to ultimate terms it appears to resolve itself into a complex of relations. And further these relations cannot be entirely devoid of comparability; for if nothing in the world is comparable with anything else, all parts of it are alike in their unlikeness, and there cannot be even the rudiments of a structure.

The axiom of parallel displacement is the expression of this comparability, and the comparability postulated seems to be almost the minimum conceivable.

But later in the same section Eddington moves on to a slightly less optimistic position, which leads to considerations of the greatest importance:

There is a certain hiatus in the arguments of the relativity theory which has never been thoroughly explored. We refer all phenomena to a system of coordinates; but do not explain how a system of coordinates (a method of numbering events for identification) is to be found in the first instance. It may be asked, What does it matter how it is found, since the coordinate-system fortunately is entirely arbitrary in the relativity theory? But the arbitrariness of the coordinate-system is limited. We may apply any continuous transformation; but our theory does not contemplate a discontinuous transformation of coordinates, such as would correspond to a re-shuffling of the points of the continuum. There is something corresponding to an *order of enumeration* of the points which we desire to preserve when we limit the changes of coordinates to continuous transformations.

It seems clear that this order which we feel it necessary to preserve must be a structural order of the points, i.e. an order determined by their mutual relations in the world-structure. Otherwise the tensors which represent structural features, and have therefore a possible physical significance, will become discontinuous with respect to the coordinate description of the world.

He then reviews Robb's attempt (Robb, 1914) to provide something of this kind in the case of special relativity but considers it too complex to provide hope of its extension to the general theory. He returns to his view, expressed in the earlier *Mind* paper, of the possibility that atomic phenomena may provide the only example of the application of true natural laws, as distinct from those originating in the mind:

The hiatus probably indicates something more than a temporary weakness of the rigorous deduction. It means that space and time are only approximate conceptions, which must ultimately give way to a more general conception of the ordering of events in nature not expressible in terms of a fourfold coordinate-system. It is in this direction that some physicists hope to find a solution of the contradictions of the quantum theory. It is a fallacy to think that the conception of location in space-time based on the observation of large-scale phenomena can be applied unmodified to the happenings which involve only a small number of quanta. Assuming that this is the right solution it is useless to look for any means of introducing quantum phenomena into the later formulae of our theory; these phenomena have been excluded at the outset by the adoption of a coordinate frame of reference.

This last sentence has to be read in the light of the ideas discussed above of descriptive tolerance and of theory-languages. So the later steps in the paths of understanding nature are not to be seen as rejecting the coordinate-frame but as keeping it alongside some other theory-language.

7   The principle of identification

The remaining two aspects of Eddington's thought are less easy to differentiate from those of other scientists, at least at the time of MTR. My reason for emphasising them is that they develop more strongly in Eddington's later work. The use of a principle of identification is a characteristically Eddingtonian move. I cannot, even speculatively, attribute the notion to any earlier thinker. Eddington can be seen to come to it gradually in MTR. It arises first in discussing the energy tensor on p. 119:

Appeal is now made to a Principle of Identification. Our deductive theory starts with the interval (introduced by the fundamental axiom of §1), from which the [metric] tensor is immediately obtained. By pure mathematics we derive other tensors . . . .. These constitute our world-building material; and the aim of the deductive theory is to construct from this a world which functions in the same way as the known physical world. If we succeed, mass, momentum, stress, etc. must be the vulgar names for certain analytical quantities in the deductive theory; and it is this stage of naming the analytical tensors which is reached in (54.3). If the theory provides a tensor . . ., which behaves in exactly the same way as the tensor summarising the mass, momentum and stress of matter is observed to behave, it is difficult to see how anything more could be required of it.

A footnote at this point makes Eddington's familiar caveat about atomicity; no clue has been found as to why matter as defined 'has a tendency to aggregate into atoms leaving large tracts of the world vacant.'

The principle is invoked again in the attempt (MTR, p. 146) to avoid a logical difficulty. General relativity has been set up by identifying the space-time interval with the results of certain manipulations with scales and clocks. Now the discussion of the energy tensor has meant that:

. . . this point of contact of theory and experience has passed into the background, and attention has been focussed on another opportunity of making the connection. The quantity . . . appearing in the theory is, on account of its property of conservation, now identified with matter, or rather with the mechanical abstraction of matter which comprises the measurable properties of mass, momentum and stress sufficing for all mechanical phenomena. By making the connection between mathematical theory and the actual world at this point, we obtain a great lift forward.

Having now two points of contact with the physical world, it should become possible to construct a complete cycle of reasoning. There is one chain of pure deduction passing from the mathematical interval to the mathematical energy-tensor. The other chain binds the physical manifestations of the energy-tensor and the interval; it passes from matter as now defined by the energy-tensor to the interval regarded as the result of measurements made with this matter. This discussion of this second chain still lies ahead of us.

If actual matter had no other properties save such as are implied in the functional form of [the energy-tensor], it would, I think, be impossible to make measurements with it. The property which makes it serviceable for measurement is discontinuity (not necessarily in the strict sense, but discontinuity from the macroscopic standpoint, i.e. atomicity).

The orthodox view here would be to see the existence of two points of contact as expressing a simple consistency requirement. It is worth noting that Eddington does not choose that way of putting it.

There are a number of bold uses of the principle of identification in Eddington's discussion of his unified field theory at the end of the book. For example it is argued on pp. 201–2 that because a certain vector which arises geometrically can be modified in a certain way ('gauge transformation') 'without altering the intrinsic state of the world' and because it is already known that just such a modification to the electromagnetic vector potential does not alter the electromagnetic field, the two are identified. Then again, (p. 219):

The (metric) tensor . . . which has hitherto been arbitrary, must be chosen so that the lengths of displacements agree with the lengths determined by measurements made with material and optical appliances. Any apparatus used to measure the world is itself part of the world . . .. This can only mean that the tensor . . . which defines the natural gauge is not extraneous, but is a tensor already contained in the world-geometry. Only one such tensor . . . has been found.

This time the identification provides Einstein's law of gravitation too.

A very clear statement of the principle comes on p. 222 of MTR:

In §91–93 we have developed a pure geometry, which is intended to be descriptive of the relation-structure of the world. The relation-structure presents itself in our experience as a physical world consisting of *space*, *time* and *things*. The transition from the geometrical description to the physical description can only be made by identifying the tensors which measure physical quantities with tensors occurring in the pure geometry; and we must proceed by inquiring first what experimental properties the physical tensor possesses, and then seeking a geometrical tensor which possesses these properties *by virtue of mathematical identities.*

If we can do this completely, we shall have constructed out of the primitive relation-structure a world of entities which behave in the same way and obey the same laws as the quantities recognised in physical experiments. Physical theory can scarcely go further than this. How the mind has cognisance of these quantities, and how it has woven them into its vivid picture of a perceptual world, is a problem of psychology rather than of physics.

To sum up Eddington's view, his principle of identification is a form of conventionalism. Mass is simply what behaves like mass. The critic who looks at Eddington's 'construction of matter' arguments and claims: 'Yes, you have proved that this behaves like matter but surely it may be something else which behaves in the same way' is just ruled out of court.

8   *Non-redundancy*

The remaining idea is closely related to the seventh. Eddington takes it for granted as a principle. In this he was in the vanguard for 1923, though it is now a view widely applied in physics, especially in high-energy particle theories. The principle is to the effect that everything constructed by the (correct) structural theory will be found in the world (or, putting it more strictly, to each mathematical construction there will correspond a feature of the world with which it will be identified). Now this proves not entirely true in MTR once the unified field theory has been constructed. Eddington notes the failure as a clue to future development. Thus, on p. 226:

Besides furnishing the two tensors . . . of which Einstein has made good use, our investigation has dragged up from below a certain amount of useless lumber. We have obtained a full tensor which has not been used except in the contracted form – that is to say certain components have been ignored entirely, and others have not been considered individually but as sums. Until the problem of electron-structure is more advanced it is premature to reject finally any material which could conceivably be relevant; although at present there is no special reason for anticipating that the full tensor will be helpful in constructing electrons.

At this time Eddington is still equivocal about the ontological commitment of theories. Indeed, on the next page of MTR he says:

We see that two points of view may be taken:
(1) Only those things *exist* (in the physical meaning of the word) which could be detected by conceivable experiments.
(2) We are only aware of a selection of the things which *exist* (in an extended meaning of the word), the selection being determined by the nature of the apparatus available for exploring nature.

Both principles are valuable in their respective spheres. Since every experimental procedure has meaning only as part of a theory, the first point of view is essentially that of non-redundancy. The second is more Kantian; we cannot be aware of 'things-in-themselves'. Non-redundancy can be read in either direction. Most people would regard it as a desirable criterion to which theories should strive and Eddington would have been with them. But he also came to use it the other way and this became more than the scientist's usual expectation of predictive power – that the theory will predict something new and that will then be observed. It is hard to say exactly in what way Eddington's view goes beyond this; partly in the strength with which he held it. *Every* structure that the theory predicts is expected to be identifiable.

Already by 1923, then, Eddington's philosophical mind was very far from a blank sheet. In addition to his self-avowed positions of selective subjectivism and the use of a principle of identification, and his borrowing of the idea of structure as the content of 'fundamental' physics from Russell, he held a number of views about scientific theories which he does not mention. The related notions of descriptive tolerance and of a hierarchy of theory-languages were to be the unavowed key ideas which would inform the Kantian operationalism and the Popperian falsifiability in carrying on the struggle to unlock the mysteries of the atom.

Notes

1. The Einstein tensor G^{ab} identically satisfies $G^{ab}{}_{;b} = 0$. The analogy is with the classical equation of continuity in hydrodynamics,

$$\frac{\partial m}{\partial t} + \text{div}(mv) = 0.$$

If one rewrites hydrodynamics in four-dimensional form, with a four-momentum $p^a = [mv^1, mv^2, mv^3, m]$, this becomes $p^a{}_{;a} = 0$ and the analogy is obvious.

2. Einstein's emphasis is on the geodesic equation,

$$\frac{dv^a}{ds} + \Gamma^a_{bc} v^b v^c = 0,$$

where v^a is the four-velocity, $v^a = dx^a/ds$, and he interprets the second term as a force in somewhat the Newtonian sense. At any point a freely-falling coordinate system will make $\Gamma^a_{bc} = 0$ so the particle is under no forces. But the evidence of a 'true' gravitational field is provided by the fact that this same particle cannot also see nearby particles as under no forces. If a nearby particle is at a displacement y^a from it ($y^a v_a = 0$) then, by considering two nearby situations in an obvious notation:

$$v^a ds + y^a + dy^a = y^a + (v^a + y^b v^a{}_{,b}) ds',$$

where

$$ds' = ds(1 + k_c y^c)$$

for some k_c independent of y^c. To the first order in y,

$$\frac{dy^a}{ds} = y^b v^a{}_{,b} + v^a k_c y^c.$$

But since $y^a v_a = 0$ and also $dy^a/ds = 0$ (the geodesic equation in the frame which is freely falling), we have $k_c y^c = 0$ so that $k_c = 0$. The remaining terms are not written in invariant form, but can be by defining the invariant derivative of y^a by $\delta y^a/\delta s = v^b y^a{}_{;b} = v^b y^a{}_{,b} + \Gamma^a_{bc} v^b y^c$. Then, in the freely-falling coordinates, the equation can be written

$$\frac{\delta y^a}{\delta s} = y^b v^a{}_{;b},$$

(since this is identical with the previous form when $\Gamma^a_{bc} = 0$) and in this form both sides of the equation are vector quantities. Now repeat the operation on the left-hand side to derive an 'acceleration' of one particle relative to the other:

$$\frac{\delta^2 y^a}{\delta s^2} = y^c v^b{}_{;c} v^a{}_{;b} + y^b v^c v^a{}_{;bc}.$$

By differentiating the geodesic equation in the form $v^b v^a{}_{;b} = 0$ the first term may be put in the form $- v^a_{;bc}v^b y^c$ and it is then easy to see that the equation becomes

$$\frac{\delta^2 y^a}{\delta s^2} + R^a_{bcd}v^b y^c v^d = 0,$$

where R^a_{bcd} is the RC tensor, as given in (Levi-Civita, 1926). The significance of the equation was not seen for many years until (Pirani, 1956). Eddington did not have this equation but he did see intuitively that it was R and not Γ that described the presence of a gravitational field.

3. Eddington's theory used an affine connexion Γ^a_{bc} which was not connected to the metric by the condition $g_{ab;c} = 0$ but was still constrained to be symmetric in b,c. Weyl's theory restricted the connexion still more.

4. Einstein produced several such theories but the most satisfactory theory was due to Schrödinger in 1948–9. It is available in book form (Schrödinger, 1950). This theory not only allows the connexion to be non-symmetric but also operates without the prior introduction of any metric at all. A symmetric Ricci tensor, R_{ab}, appears in it and g_{ab} is then *defined* as proportional to it. Einstein's 1916 law of gravitation $R_{ab} = k g_{ab}$ is then an identity. This way of proceeding may be compared with Eddington's discussion of matter earlier in the chapter.

5

'Something has slipped through the net'

This chapter is devoted to the one major error in MTR which I have mentioned above. Our knowledge of the external world will, it is argued there, be precisely of the form expressible by the tensor calculus. On p. 49, for example, Eddington says:

I do not think it is too extravagant to claim that the method of the tensor calculus, which presents all physical equations in a form independent of the choice of measure-code, is the only possible means of studying the conditions of the world which are at the basis of physical phenomena.

That Eddington, in common with all physicists (and most mathematicians who were interested), was so convinced was of the greatest importance on a personal level. When his error was exposed by Dirac's 1928 paper on the electron, it had a profound psychological effect. For Dirac expressed the electron equation, not in tensor form, but in terms of new entities – spinors. Since this is such a key issue for my argument, I make no apology for going into it at some length. In the course of the discussion I shall be able to take up the question, left from Chapter 4, of what exactly a tensor is.

The principle of relativity

In order to understand Eddington's conviction and also to see why it is false, it is necessary first to look more closely at the so-called 'principle of relativity' which is mentioned by both Einstein and Eddington but without saying exactly what the principle is. Einstein, for example, says (Einstein, 1916):

If a system of coordinates K is chosen so that, in relation to it, physical laws hold good in their simplest form, the *same* laws also hold good in relation to any other system of coordinates K' moving in uniform motion relative to K. This postulate we call the 'special principle of relativity'.

Later he generalises this by allowing any motion of K', so removing the restriction 'special'. But the principle as stated refers in rather general terms to 'physical laws' (and there is also a worrying qualification that they be true 'in their simplest form') and says nothing in detail about how these laws are expressed. This is strange, since the reference in the principle to the use of coordinate-systems implies that the whole statement is about means of expression.

Clarity about the principle came only with Hermann Weyl some years later. Weyl's discussion is subtle and occurs in several different forms. I use the term 'Weyl's argument' for a paraphrase which I give here, with which he might or might not have agreed. The argument falls into three parts. Part 1 formulates the criterion for a quantity measured in a reference frame (coordinate-system) to be a physical quantity. (Mathematicians will see this as a filling out in the specific physical context of Klein's *Erlangenprogramm* for geometry (Klein, 1872).) Part 2 then describes how to find sets of numbers fulfilling the criterion. Many of these sets are tensors, as Eddington knew. Then, in Part 3, I carry through the procedure for a very simple case to exhibit the non-tensor quantities ('spinors') which also fulfil the criterion.

PART 1   Weyl's argument

The necessity for the principle of relativity, as explained above, arises in the use of reference frames. A passenger travelling in a train moving at 100 miles per hour will, if provided with a thermometer, determine the same air temperature as another observer at rest on the ground. But if a road is running beside the railway track and the ground observer is passed by a car going in the same direction as the train but at only 70 miles per hour, the train passenger will see the car as having a speed of $70 - 100$, -30, that is, 30 miles per hour in the opposite direction. Some quantities, then, like temperature, are independent of the reference frame, and others, like speed,

are not. The first set of quantities are evidently excellent candidates for
physical quantities, describing 'conditions of the world'. Above the second
set hangs a question mark. If they are different in different reference frames,
is this perhaps because they are purely artefacts of the reference frame,
telling us nothing about the conditions of the world? For example, if the
train is suddenly slowed by a signal and the passenger's cup of coffee slops
over into the saucer, the effect he observes does not correspond to anything
external in the physical world, but simply to the sudden change in motion of
the reference frame. (By using the train as a physical representation of his
reference frame I have somewhat blunted the sharp distinction. If the train
is to be employed in this way, then one must take 'the physical world' as
including everything except the train.) Can there perhaps be an intermediate type of quantity, different for different coordinate-systems and yet able
to tell us something about the physical world? Indeed there can, as the
example about car speed shows and it transpires that the most important
quantities are of this type. The tensors, of which Eddington is so fond, are
such quantities but, contrary to his beliefs not the only ones. To see both the
reasons for his beliefs and how they can be false it is necessary to follow
Weyl's argument in more detail.

Implicit in Weyl's argument is the idea that the problem is not simply to
find the criterion for determining what we may call, for shortness, physical
quantities. By this term I mean numbers or sets of numbers, that may have a
different value or values in different reference frames but are not mere
artefacts of the reference frame, since they tell us something about the
'condition of the world'. Rather, according to Weyl, the problem is to define
more carefully what a physical quantity should be. When we have done
that, the criterion will arise naturally. Weyl investigates this by considering
three reference frames; call them *A*, *B*, and *C*. There will then be
transformations between these, which allow us to get from the coordinates
and time of any event according to one reference frame to the corresponding coordinates and time according to the other. For example, as far as the
spatial coordinate-system is concerned, the three directions along which
the coordinates are measured might be chosen in different directions in the
different frames. The transformations will then be rotations. Again,
different zeros of time-reckoning might be chosen, or different units of time
(seconds or hours). The transformations will then be of the form

$$t \rightarrow t' = Pt + Q$$

where P,Q are constant numbers and P is positive (so that the time runs the same way in both reference frames). More importantly, if there is a case in which one reference frame is moving relative to another in special relativity, the transformations will be Lorentz transformations. In general relativity 'any coordinates may be used'. This has to be taken with a pinch of salt. Taken literally, it would mean that the coordinates and time corresponding to an event in frame *B* could depend on those of the same event in frame *A* in 'any way'. But in the mathematical formulation of the theory this idea of 'any' is restricted by conditions of smoothness so that two nearby events which have roughly the same coordinates and time in *A* have the same property in *B*. In any general case, I shall use the name $T(A,B)$ for the transformation which converts coordinates and time in *A* to those in *B*. Thus $T(A,B)$ is a rule which will tell me the values of the coordinates and time of an event in frame *B* given them in frame *A*. In the example above (of change of time zero and unit) the equation

$$t' = Pt + Q$$

is what is denoted by the rule $T(A,B)$.

So much for the changes in coordinates and time. Now suppose that a proposal is made about a physical quantity in frame *A*. This proposal is that a certain number, or collection of numbers, represents the quantity. I say a collection of numbers to cover the cases like that of speed, where in the three-dimensional case one would either specify the magnitude of the speed and then its direction separately or give three components of speed (northwards, eastwards and upwards for instance) in three different directions. I shall denote the number, or the collection of numbers, in frame *A* by *a*. If these numbers are proposed as physical quantities, there must be corresponding ones in frame *B*, and if the collection in *A* has more than one member, then the collection in *B* must have the same number. I call the collection in *B,b*.

When the transformation $T(A,B)$ is performed to change the space and time numbers of an event in frame *A* to those in frame *B*, there will be corresponding changes in the numbers *a* which will turn them into the numbers *b*. It may not be clear, without careful investigation, what these changes are. I shall suppose that such an investigation has been carried out and the result of this is to give the rules to be applied to *a* to derive *b*. I call this rule $t(A,B)$. To make this clear, consider the case of a moving frame of reference in Newtonian mechanics, with a three-dimensional set of space coordinates (x,y,z) and a time t. Suppose that one frame is moving with constant speed v along the x-axis of the other. The rule $T(A,B)$ is then

$$x' = x - vt, y' = y, z' = z, t' = t.$$

If now a is taken as the set of three numbers representing the velocity of a particle, say $a = (p,q,r)$ where these are the components of velocity along the three coordinate directions, and correspondingly $b = (p',q',r')$, the rule $t(A,B)$ is

$$p' = p - v, q' = q, r' = r.$$

This rule may be found by drawing a diagram and arguing geometrically, or by differentiating the rule T. In either case, an investigation is needed.

Exactly the same considerations apply when this change of reference frame is followed by a further one from B to C by means of the rule $T(B,C)$; and there will be a corresponding rule $t(B,C)$ taking the set b into the set c. Now, however, it is clear that the effect on the coordinates and times in A of $T(A,B)$ followed by $T(B,C)$ is simply to produce the coordinates and time in C; in other words, to execute the rule which I have called $T(A,C)$. The T transformations have in this way a kind of transitivity. The resultant of doing the two transformations in succession is the same as doing another transformation of the same kind. In mathematician's jargon this is expressed by saying that the whole set of transformations forms a *group*. (So a technical meaning is added to the normal English connotation.) In special relativity the discussion centres round the *Lorentz group*. In general relativity the more general transformations, restricted only by smoothness, constitute the *Einstein group*. In the simpler case I mentioned above, when the spatial coordinate-axes are differently oriented in the two reference frames, the *rotation group* is involved.

But what of the corresponding transformation rules for the sets of numbers claiming to represent a physical quantity? When $T(A,C)$ is carried out, there will be a transformation rule $t(A,C)$ giving the numbers in the set c in terms of those of set a. Will this likewise be the result of first performing $t(A,B)$ and then, on the numbers b which result, performing $t(B,C)$? That is, will the t transformations have the same kind of transitivity? Weyl's answer is that the condition that this should be so is the required definition for the sets a, b, c to be the measures of a physical quantity referred to each of the reference frames A, B, C. For this requirement simply states that it does not matter, in going from a in A to c in C whether a call has been made at frame B on the way, or not. If the numbers, instead of being physical quantities, were mere artefacts of the reference frame, then, of course, the introduction of the extra reference frame B will make a difference. The criterion is, then, that whenever $T(A,B)$ followed by $T(B,C)$ is $T(A,C)$ – as it will always be here because of the notation I have adopted – then also $t(A,B)$ followed by $t(B,C)$ will produce the same result as $t(A,C)$. The mathematician's jargon for this is to say that the ts are a *representation* of the group of the Ts.

PART 2   Finding representations

The next step is to try to discover all the representations of the group of Ts. In this way all the possible forms of physical quantity would be exhibited; whether these forms are all found in the external world is another matter, to be determined by experiment. This has to be done afresh for each group of Ts but there is a useful device for every physical group which helps the investigator in knowing where to look. The device is this:

(a)   first think of some set of numbers which one can be pretty sure represent a physical quantity;

(b)   then verify this by finding how the set of numbers is transformed and checking that the rules do indeed form a representation of the group of Ts;

(c)   if this is the case, then a representation of the group has been found.

Any set of quantities which transform under the representation will also be a set of physical quantities; one, indeed, transforming just like the set chosen in (a). This technique seems deceptively simple but it is very effective. It even includes the production of a representation for quantities (temperature was the prototype in the train example) which do not change when the reference frame is changed. The representation in this case is an easy one to spot. For each $T(A,B)$ it is required to specify the rule $t(A,B)$; this rule, for quantities like temperature, is simply $b = a$. Similarly for $t(B,C)$ the rule is $c = b$. Now when $b = a$ and $c = b$ it follows that $c = a$; this should be the rule $t(A,C)$ and indeed it is, so that the specification of t is a representation of the group. It is noteworthy here that nothing has been used in the argument about the transformations T, so that exactly the same argument applies to any group. The quantities 'transforming under this representation' (that is, staying unchanged) are called *invariants* or *scalars*. The mathematicians refer to this representation somewhat slightingly as the *trivial representation*; it is a representation of any group.

A more detailed application of the device is to take for a the set of four numbers corresponding to a set of very small changes (displacements) in the coordinates and time. By 'very small' here is meant that the formulae for the corresponding numbers in the second reference frame cannot involve squares or products of the quantities, since these would be even smaller, and so each number in the second reference frame must be simply a sum of multiples of the numbers in the first. In mathematician's language, the transformation t will be linear. The ts do form a representation of the Einstein group.[1] This representation is known as that of contravariant vectors. The reason for the adjective is that another representation, called that of covariant vectors, also arises. This is that constructed from the transformation of the four rates-of-change of an invariant function of position with respect to time and space. It will be recalled that such a collection of rates-of-change appeared above in the discussion of Newtonian gravitation and Laplace's equation, though in that case the number of dimensions was three and not four. This use of the terms contravariant and covariant is, in fact, in agreement with their use for coordinates in Chapter 2.[2] In this way are derived the two basic *vector representations*. The detailed mathematics of these representations need not delay us here.

The important point is that from these last two representations it is now easy to derive an unlimited number. This is done in the following way: take any number of vectors (in the sense of the word just described); to fix ideas let it be one contravariant vector v and one covariant vector w. (Any other collection would do equally well.) Physically it might be, for example, that v was a velocity and w a position. Now construct the set S of numbers in this way: since v is a vector it is shorthand for four numbers. I call these v_1, v_2, v_3, v_4. Similarly for w. The set S consists of all products like $v_3 w_2$ and so on; that is, a collection of $4 \times 4 = 16$ numbers. Because v,w are vectors and, in going from reference frame A to frame B, each vector of the pair transforms according to the rule given, it is easy to see that S transforms under a representation of the group and so is a physical quantity. The representations constructed in this way are called the *tensor representations*. The quantities transforming under them are called *tensors*. When needed, one can speak of tensors of rank 2 to mean that, as here, two vectors are involved in the construction. Then vectors themselves are tensors of rank 1. One can also if need be speak independently of the contravariant and covariant ranks.[3]

Tensors do not have the easy intuitive clarity of vectors. The name arises because they were recognised first as of importance in elasticity but that is not relevant here. It is best not to seek for geometrical pictures but to concentrate on the notion of a set of numbers.

Undoubtedly part of Eddington's enthusiasm for tensors, and his conviction that here there is everything needed to describe the 'condition of the world' simply rested on the prolific character of this generation of representations, though he would not have expressed it in that way. Some people would still argue that, because of the generality of the Einstein group, only tensor representations will ever be needed in general relativity. Really, however, they are arguing that it is useful to restrict the definition of what constitutes general relativity in this way, so that the point is only a semantic one. There is more to be said in favour of the view that Eddington expressed. It is true that tensors arise very naturally in the course of doing theoretical physics. The basic operation of multiplication of entities is important in itself but it is also a kind of prototype for other more recondite operations (differentiation and so on). If one begins with the vector representations, one will not be led outside. The question is where one should begin.

PART 3   Spinors

To answer this question without too much technical detail, I shall look at a tiny fragment of the Einstein group. This fragment consists of the rotations implied by the arbitrary choice of *two* of the three spatial directions, say those of x,y leaving the z-direction fixed, as well as the time. Different reference frames A,B will then be distinguished by the two directions for B both being rotated through a certain angle R from their positions in A. This specifies the rule $T(A,B)$ completely.[4] The whole group of Ts is the assemblage of all such rules for different values of R. The resultant transformation of rotation through R followed by rotation through S is evidently rotation through $R + S$. This exhibits the group property of the transformations, as defined above. The vector representations of this group are easily found; for the small displacements, from the transformation of which the contravariant vector representation is constructed, transform exactly like the coordinates themselves, so that $t(A,B)$ is the same as $T(A,B)$. It turns out that the same is true for the covariant vector representation.

When the tensor representations of rank two are considered there is a surprise. To make this surprise clear, I think it is best to look at the formulae in a little more detail. I give here a very straightforward way of looking at it. The mathematician may prefer that in the notes.[5] From Fig. 5.1 the original rotation through R has the form (writing x^1,x^2 instead of x,y)

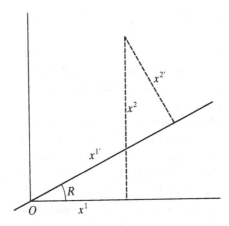

Fig. 5.1. Changes in coordinates under rotation through R.

$$x^{1'} = x^1\cos R + x^2\sin R,$$
$$x^{2'} = -x^1\sin R + x^2\cos R.$$

A tensor of rank two would be a set of numbers like T^{ij} transforming like $v^i w^j$ where v,w are vectors behaving, as we said, just like the coordinates. For example,

$$v^{1'}w^{1'} = (v^1 c + v^2 s)(w^1 c + w^2 s), \qquad (c = \cos R,\; s = \sin R)$$
$$= v^1 w^1 c^2 + (v^1 w^2 + v^2 w^1)sc + v^2 w^2 s^2.$$

Accordingly

$$T^{1'1'} = T^{11}c^2 + (T^{12} + T^{21})sc + T^{22}s^2.$$

In the same way

$$T^{2'2'} = T^{11}s^2 - (T^{12} + T^{21})sc + T^{22}c^2.$$

Finally

$$T^{1'2'} = -T^{11}cs + T^{12}c^2 - T^{21}s^2 + T^{22}sc,$$
$$T^{2'1'} = -T^{11}cs - T^{12}s^2 + T^{21}c^2 + T^{22}sc.$$

Since $\cos^2 R - \sin^2 R = \cos 2R$ and $2\sin R\cos R = \sin 2R$, the surprise is evident. $T^{11} + T^{22}$ and $T^{12} - T^{21}$ are invariant and can be put on one side for the moment. The pair of numbers $(T^{11} - T^{22}, T^{12} + T^{21})$ behave exactly like two component parts of a vector with one exception: the rotation is through the double angle, $2R$. Moreover, the four numbers just listed are such that they uniquely define the original four members T^{ij} of the set S.

The mathematically inclined reader may look in vain amongst the higher rank tensor representations for such a simple breakdown. To summarise the position that the argument has reached: the rank two tensor representations, and only those, of the rotation group in two dimensions have an extremely simple decomposition in which two of the four numbers transform just like vectors but through double the angle. It is obvious that the rule 'Rotate through double the angle' is a representation of the rotation group because if $R + S = T$, then $2R + 2S = 2(R + S) = 2T$.

This curious state of affairs is open to a quite different and highly instructive interpretation. This new way of looking at the formulae is to imagine the *new* quantities to be the ones defining the vector representation and so also the coordinate transformation itself. This means that the angle of rotation is $2R$, which I will call S. The argument can then be reversed so as to show that such a vector can be written as sums of certain products of other quantities, ϕ say, the numbers which in the old interpretation were the components of the vectors. Thus, if ϕ,ψ are two such quantities, then $(\phi^1\psi^1 - \phi^2\psi^2, \phi^1\psi^2 + \phi^2\psi^1)$ will be a vector and $\phi^1\psi^2 - \phi^2\psi^1$ and $\phi^1\psi^1 + \phi^2\psi^2$ will be invariants. If now S is the defining rotation angle, these new ϕ quantities will transform under a rule like that of vectors but through an angle $R = \frac{1}{2}S$. It is clear from the method of construction that this is a representation of the group. Indeed, the same can obviously be said of transformations which are through any multiple of the original angle. But the new representation is of a different form from the tensor representations. This may be seen by considering two coordinate rotations, one through S and one through $S + 360°$. These two rotations have exactly the same effect on the coordinates, since $360°$ is a complete turn. But the new quantities, ϕ, are rotated in the two cases by $\frac{1}{2}S$ and $\frac{1}{2}S + 180°$ respectively. The result of the second transformation is therefore different from that of the first. In fact, it produces quantities of the same magnitude but opposite sign. The new representation is said to be a *two-valued representation* because two formulae $t(A,B)$ correspond to a single $T(A,B)$. The quantities, ϕ, transforming under the two-valued representation are called *spinors*.

If rotation through $\frac{1}{2}S$ is a representation of the rotation group, so also is rotation through $\frac{1}{3}S$, $\frac{2}{3}S$, $\frac{1}{4}S$ and so on; so there are also three-valued, four-valued etc. representations. These are of no physical importance whereas the two-valued ones are. The reason for this is as follows: the rotations in the plane are, of course, but a part of the whole rotation group in three dimensions. The same considerations about spinors will apply in the larger group. This larger group has a certain complication, which is usually expressed by saying that it is not simply-connected. The reader may illustrate this fact, which is not at all a trivial one, by a little experiment. Cut a long narrow ribbon of paper of length at least twenty times its width and mark a cross at each end on the same side. Hold one end of the paper in each hand and twist one end through a complete turn so that, when the two crosses face you, the ribbon is straight but not flat – it has a screw form. By moving the crosses about in space (always keeping them facing you) you can make the ribbon flat at the expense of its being twisted into part of a ring, but any attempt to make it straight again will cause the screw shape to reappear. Now repeat the experiment but twisting through two complete turns at the beginning. The screw shape is now more pronounced, but you will find that you will be able to make the screw disappear by moving the ends about whilst keeping the crosses facing you. This is a situation, then, in which a twist through two complete turns is not the same as through one. But a twist through three complete turns is the same as through one; and so on. The mathematicians summarise this situation by saying that the rotation group is doubly-connected. It is this that makes the two-valued representations physically important, but not the other multiple-valued ones.

None the less there is a further query about the spinors. They fulfil the Weyl condition about physical quantities but only in a modified way *via* the use of two-valued representations. This introduces an essential ambiguity into their values. The result of rotating the coordinates through 360° – which produces the identical position of the coordinate-system – changes the sign of the spinors. They may well be able to arise during physical theorising and they are not mere artefacts of the reference frame; but they cannot directly represent the measured value of a physical quantity, like temperature, with a unique value. Only the square of a spinor, or some more complicated product involving, amongst other things, two factors, each the same spinor, could be such a quantity. But it will be noted that Eddington, in the quotation at the beginning of this chapter, did not restrict his claim about tensors to quantities which could be directly measured, but to 'physical equations'. Spinors serve to show the incorrectness of his claim.

This concludes the third and final part of the argument. It is sufficient to show that it is possible to have physical quantities, in the sense defined, transforming under two-valued representations. It cannot show, of course,

that such quantities really will occur in physics. If Eddington could have
been aware of Weyl's argument (in fact it was not published until well after
MTR), his own non-redundancy principle would have convinced him of the
physical significance of them. But, in common with the whole community of
physicists, he was unaware of the theoretical existence of spinors until 1928,
although the necessary mathematics was known (Cartan, 1913) in the
context of projective geometry.[6] There was another clue from 1926
onwards for, in Schrödinger's version of the new quantum theory of that
year there arose a quantity ψ whose value could not be determined by
experiment. It was not a spinor, for Schrödinger's theory was not
relativistic; indeed the transformation of his equation was then a muddle.[7]
But ψ had the property that the square of its magnitude had a physical
interpretation. This clue was of no use to Eddington. In the first place, he
had no great interest in the technical aspects of the new quantum theory,
although he accepted the ideas of the theory readily. But secondly the
question of changes in reference frame was never tackled in any clear
manner by the workers in the theory. None the less, it was from quantum
mechanics that the great upset of Eddington's ideas was to come, and it is to
the development of that I now turn.

Notes

1. Take $x^1 = x$, $x^2 = y$, $x^3 = z$, $x^4 = t$. The set a consists of (dx^c) $(c = 1, \ldots 4)$.
   Since $x^{c'}$, the coordinates in B, are smooth (here, at least once-differentiable)
   functions of those in A, we have:

   $$dx^{c'} = X^{c'}_c dx^c, \text{ where } X^{c'}_c = \partial x^{c'}/\partial x^c.$$

   This is, as expected, a representation because, if

   $$dx^{c''} = X^{c''}_{c'} dx^{c'} = X^{c''}_{c'} X^{c'}_c dx^c,$$

   then

   $$X^{c''}_{c'} X^{c'}_c = X^{c''}_c,$$

   by the chain rule in partial differentiation. Similarly, the prototypes of covariant
   vectors are derived by noting that

   $$\phi_{,c'} = X^c_{c'} \phi_{,c}$$

   and the chain rule again proves this to be a representation.
2. In Fig. 2.1, and with the more usual notation for contravariant and covariant
   vectors, write $x = x^1$, $y = x^2$, $X = x_1$, $Y = x_2$. Since

   $$ds^2 = (dx^1)^2 + (dx^2)^2 + 2\cos\alpha\,dx^1 dx^2,$$

the metric form has $g_{11} = g_{22} = 1, g_{12} = \cos \alpha$. Hence

$$x_1 = g_{11}x^1 + g_{12}x^2 = x^1 + x^2\cos \alpha$$

and this agrees with the geometry. Similarly for x_2.

3. So here a typical tensor representation is constructed by

$$u^{a'}v_{b'}w_{c'}x_{d'} = X^{a'}_a X^b_{b'} X^c_{c'} X^d_{d'} u^a v_b w_c x_d$$

and the fact that this is a representation again follows from the chain rule. A general tensor of this type, $C^a{}_{bcd}$, for which

$$C^{a'}{}_{b'c'd'} = X^{a'}_a X^b_{b'} X^c_{c'} X^d_{d'} C^a{}_{bcd},$$

is not of the form of a product but it can always be written as a sum of a suitable number of such products.

4. The usual coordinate transformation is

$$x' = x\cos R + y\sin R,$$
$$y' = -x\sin R + y\cos R,$$

which may be more conveniently written as

$$x' + iy' = (x + iy)e^{-iR}.$$

5. Let the two vectors in the product be (x,y) and (u,v) so we have to consider the set $(xu,xv,yu,yv) = (A,B,C,D)$ say. Then

$$(x + iy)(u + iv) = A - D + i(B + C)$$

so that

$$(A' - D') + i(B' + C') = [(A - D) + i(B + C)]e^{-2R},$$

whilst

$$(x + iy)(u - iv) = A + D - i(B - C)$$

so that

$$A' + D' - i(B' - C') = A + D - i(B - C).$$

This shows that $A + D, B - C$ are invariants and that $(A - D, B + C)$ transforms like a vector but rotated through double the angle. The same technique will show that the eight components of a rank three tensor fall into four pairs. One pair rotates like a vector but through $3R$ and the others all rotate through R. With rank four tensors there are six invariants and the remaining ten components split into five pairs, one rotating through $4R$ and the remainder through $2R$.

6. In fact the basic idea has been around, somewhat unappreciated, for nearly two centuries in the form of Euler's parameters in rigid dynamics. A very near miss to a complete formulation comes in 1843 with Hamilton's discovery of quaternion algebra. In a modern notation, take e_i $(i = 1,2,3)$ as unit vectors along three rectangular axes and define a product between them by

$$e_i e_j = \varepsilon_{ijk}e_k - \delta_{ij},$$

where ε_{ijk} is the alternating symbol, $= 1$ if i,j,k is an even permutation of 1,2,3, $= -1$ if it is an odd one and $= 0$ if it is not a permutation and δ_{ij} is the Kronecker delta, $= 1$ if $i = j$ and 0 otherwise. It may be immediately verified that the product is associative, that is,

$$e_i(e_je_k) = (e_ie_j)e_k$$

and also that the product of two vectors u,v is expressed in terms of the so-called 'products' of vector analysis by

$$uv = u \wedge v - u{\cdot}v.$$

The quaternion $q = \cos\theta + u\sin\theta$, where u is some unit vector, produces a rotation, of axis u, through 2θ, by means of the equation

$$v \to v' = qvq^{-1}.$$

(Here q^{-1} is easily seen to be $\cos\theta - u\sin\theta$). This is easily verified in the case when $u = e_3$ and the general case follows. Increasing θ by π changes the sign of q but leaves the rotation unchanged. Such a formulation of rotations points to quantities transforming under

$$s \to s' = qs$$

and the ambiguity in q makes this a two-valued representation. A quaternion like s is essentially a pair of spinors, since s has four components.

7. The matter is now clear. Schrödinger's equation for a free particle of mass m comes from 'quantising' the energy equation $\mathbf{p}^2/2m = E$ and so is

$$\nabla^2\psi + (2mi/\hbar)\frac{\partial\psi}{\partial t} = 0.$$

If $\mathbf{r}' = \mathbf{r} - \mathbf{V}t, t' = t$, then $\nabla' = \nabla$, and $\partial/\partial t' = \partial/\partial t + \mathbf{V}{\cdot}\nabla$, so that

$$\nabla^2\psi' + (2mi/\hbar)\left(\frac{\partial\psi'}{\partial t} + \mathbf{V}{\cdot}\nabla\psi'\right) = 0$$

must be equivalent to the original equation. But ψ, despite Schrödinger's Herculean efforts to the contrary, has to be taken as complex, though the theory is independent of its phase, at least as far as instantaneous measurements are concerned. (Schrödinger says in a letter to Lorentz, 6 June 1926, 'What is unpleasant here and certainly to be objected to, is the use of complex numbers. ψ is surely fundamentally a real function . . .'.) Accordingly, $\psi' = e^{if}\psi$ and ψ describe the same state if f is real. Substituting this in the transformed equation and using the original one easily gives the conditions:

$$\nabla f = -m\mathbf{V}/\hbar, \ \partial f/\partial t = \tfrac{1}{2}mV^2/\hbar$$

which integrate to give the phase change $f = \tfrac{1}{2}m(V^2t - 2\mathbf{V}{\cdot}\mathbf{r})/\hbar$ to establish the required invariance. It is easy to verify that this transformation of ψ is one under a representation of the Galilean group.

6

Quantum mechanics

I conclude the first part of this book by carrying the story up to 1928, to the lightning flash in Eddington's mind produced by Dirac's paper on the wave equation of the electron. Setting out the context and Eddington's thinking about it is a very different matter for quantum mechanics and relativity. Both theories are commonly expressed in austere mathematical language. In the case of relativity the basic ideas behind the mathematics are now well understood and can be set out with little complication. These ideas were mostly already clear to Eddington. But for quantum mechanics the mathematics, yoked to the wealth of experimental results, drove the development of the theory at a breakneck speed, mostly without anyone pausing for deeper understanding. It is indeed only in the last decade that it has become generally accepted that quantum mechanics still lacks any coherent interpretation. To explain Eddington's context it is therefore necessary to say something about the mathematical formalism. It is true, as will become clear, that Eddington was a little more sceptical than most, but by and large the mathematics drove him as it did those directly working in the theory.

I begin, then, with a description of that part of the early (pre-1925) history of quantum mechanics that was in Eddington's mind in 1923. Then I explain what happened in 1925–6 and Eddington's reaction to it. The chapter concludes with Dirac's relativistic wave equation of 1928 and Eddington's further reaction to that.

The old quantum theory

What is now called the old quantum theory arose at the turn of the century. Two related experimental contradictions with classical physics played important roles in this. Maxwell's electrodynamical equations give rise to wave-like solutions, interpreted as light or other radiation (heat, X-rays and so on). The analogy here is with water-waves, though in some ways this analogy is not a good one since the 'medium' of the electromagnetic waves is the (non-existent) aether. A water-wave has a *wavelength*, *L*, the distance between successive crests and a *frequency*, *f*, the number of crests passing a point in unit time; the speed of advance of the wave is then distance travelled per unit time, that is, *Lf*. The wavelength of a light-wave is less easy to visualise, though it is well defined, but the frequency is intuitively clear and since $Lf = c$, where *c* is the speed of light, this can be used to define a wavelength. The different forms of radiation simply correspond to different values of *f*. One of the two classical contradictions concerned the equilibrium of radiation and matter. A metal poker put into the fire for a short time glows a dull red. Put deeper into the fire and for a longer period, so that it gets hotter, it becomes a brighter red and ultimately a whitish colour. So there is a correlation with the temperature *T* of matter and the frequency distribution (determining the colour of the radiation) which is in equilibrium with it. Because the laboratory experiments considered the radiation in a heated metal box, and the theoretical work dealt with the matter as a 'black body', that is, one which absorbs all the radiation falling on it, the problem was referred to as that of cavity radiation or black-body radiation. There were two formulae proposed for the relative amounts of radiation at different frequencies at a given temperature. They were derived in different ways but neither was in agreement with experiment. One of them even gave the total energy as infinite.[1] Historically it was this contradiction that gave rise to the start of the quantum theory, but it could equally have been the other described below.

The second, related, contradiction can be exhibited by striking a match. The flame is of a predominantly yellow hue. The light can be passed through a prism to break it up into its constituent colours. The result is surprising. Instead of a gradation of colouring centred on the yellow part of the spectrum and tailing off on either side, a number of sharp spectral lines, that is, colours of very precise frequency, are seen. The strongest are two yellow lines, which explains the general appearance of the flame. A more refined version of this experiment is the chemist's flame test, where a platinum wire is dipped into a material in powder form to be analysed. The wire is then heated by a Bunsen burner (i.e., a colourless flame). Sharp spectral lines appear, characteristic of the elements of the material being analysed. If the test is carried out carelessly, with the wire not clean, the principal lines are again the yellow ones noted with the match. This is characteristic of sodium, which is widely present in impurities of all kinds. To call the existence of sharp spectral lines a contradiction with classical physics is to leap ahead. The nineteenth century chemists simply saw them as a convenient observed fact.

The old quantum theory began in 1899 when Planck took up the problem of black-body radiation. His first move was to interpolate between the two formulae already derived. One of these, due to Rayleigh, had the form $f^3 \cdot (1/x)$, where x is written for f/T, for the energy density at frequency f. The other, experimentally derived by Wien, was of the form $f^3 \cdot e^{-bx}$. Rayleigh's law was good for small x, and Wien's for large. Planck's argument was somewhat roundabout, but the problem is really to find an expression in x which is like $1/x$ when x is small and like e^{-bx} when x is large. Evidently $1/(e^{bx} - 1)$ is likely to be right. In this way Planck derived a formula in good agreement with measurements.

In the latter part of 1900 Planck strove for some theoretical understanding of his formula. He tried a different approach. He looked on the calculation of an average energy (at a particular frequency) as a statistical problem, that of distributing certain numbers of 'packets' of energy, each of amount E. Planck's idea was, after the calculation, to let E shrink to a vanishingly small value, so that the steps of energy produced by the separate packets would be insignificant. Most variables in physics like time or temperature are continuous, but there are also discrete variables, like the number of a particles in a box. These are less common, and prior to 1900 very much less so. Planck started out with the idea of a continuous energy but he found it necessary, in order to derive his formula, to make the size of the packets of energy proportional to the frequency of the radiation, all the packets of one frequency being of one size.[2] The packets of energy, or 'quanta' were then of magnitude hf, where h was a new physical constant, now called Planck's constant. I described this as a contradiction because this discrete energy has to be imported, as it were, into a whole complex of theories (Maxwell's electrodynamics, thermodynamics) set up on the implied assumption of continuous energy.

It is an obvious exaggeration to refer to the *ad hoc* rule, $E = hf$, as a theory, but scientific developments often begin with an *ad hoc* rule, or even an inconsistency, from which a rational theory grows. It was not so with the old quantum theory. Planck's assumption by itself was of limited power, though in Einstein's hands it explained the curious characteristics of the photo-electric effect (the generation of electricity when light falls on certain metals). Hallwachs had found in 1888 that high-frequency light falling on suitable metals generated a current (released electrons from the surface). The energy of the electrons released did not, strangely, depend on the intensity of the light but only on its frequency. According to Einstein this was because the energy of each released electron was simply the excess of hf over its binding energy. The role of the intensity of the light was to determine the number of electrons released.

It was experimental discoveries about atomic structure that allowed an explanation of sharp spectral lines by augmenting Planck's assumption with another one. The hydrogen atom, to take the simplest case, was envisaged as a miniature solar system with the proton corresponding to the sun and a single electron as the planet in orbit round it, the force between them being electrostatic rather than gravitational. In 1913 Bohr noticed that this model could explain the sharp spectral lines of hydrogen very accurately if one assumed that only certain orbits were 'allowed' in the sense that the electron could move in them without any radiation being emitted.[3] This was a confrontation with Maxwell's theory, according to which any accelerated charge must radiate. When the electron 'jumped' from one allowed orbit to another of lower energy, the difference energy, E, was given off as radiation of a sharply determined frequency f, determined by Planck's $E = hf$. This contradiction with Maxwell's theory came to be accepted quickly. The reason for this is probably that the rule given by Bohr to determine the allowed orbits involved a new physical constant, and this new constant was almost exactly Planck's again. In fact, it was $h/2\pi$, and this constant reappears so often that it has received the name $\hbar$. The radius of the nth allowed orbit then turns out to be $n^2\hbar^2/me^2$. The integer n is a *quantum number*. This appearance of $\hbar$ seemed to give a modest coherence to the proliferation of *ad hoc* rules.

Difficulties of the old quantum theory

For the next decade quantum theory was obsessed with spectroscopy – with the effort to repeat the hydrogen success with other atoms. The problem was to know what were the right integral relations. For the alkali atoms, for example, it was argued that the influence of the nucleus on one electron was partly screened off by the other electrons so that two non-integral parameters were needed. But more complex cases could be explained only by means of extra rules and gradually a number of cases collected which no amount of ingenuity was able to subsume under the general scheme. The spectrum of neutral helium, for example, was completely at variance with the predicted values.

I think it is this generally chaotic and muddled situation in quantum mechanics which led Eddington to believe that it was, perhaps, here that the true physical laws were to be sought. The very tidiness of relativity was an indication that it was a theory of structure alone. In what other ways did this affect Eddington's thinking? We are fortunate in having his 1927 view of quantum mechanics set out in a popular book (Eddington, 1928a). This was the book form of the Gifford lectures given in Edinburgh in the previous year. But there is more to the book than this. There had been a long-fought rivalry between Jeans and Eddington, particularly over speculations in astrophysics. This was at the time a field that lent itself very well to polemical destruction of opponent's views. One late afternoon in the 1920s Eddington came down the steps of the Royal Astronomical Society's rooms in Piccadilly and met J. G. Crowther. 'Crowther' exclaimed Eddington, 'is it true that Jeans has written a book that sold 50 000 copies?'. When Crowther said that this was true, Eddington returned 'I shall write one that sells 100 000.' It was not only the edited Gifford lectures that sold so spectacularly. The same was true of *The Expanding Universe* in 1933 and of *New Pathways in Science* two years later. Eddington became one of the greatest of popular scientific writers of the time. There was a price to be paid. The essence of popular scientific writing is to tell less than the whole truth. The good writers, like Eddington, are those who contrive to do this without making actually false statements, and who none the less carry the reader along with the feeling that nothing of importance has been left out. I shall argue later that Eddington's acquisition of this skill to a high degree is one cause of the obscurities of FT.

Eddington's chapter on the old quantum theory (Eddington 1928a) shows how very well he had appreciated the differences to which it must give rise. He gives a general description of it, concentrating on the way in which Planck's constant keeps arising. Then he concludes with Bohr's 'correspondence principle':

The classical laws are the limit to which the quantum laws tend when states of very high quantum number are concerned.

He goes on to draw the conclusion:

We must not try to build up from classical conceptions . . . . We must start from new conceptions appropriate to low as well as to high numbered states . . . I cannot foretell the result of this remodelling, but presumably room must be found for a conception of 'states', the unity of a state replacing the kind of tie expressed by classical forces. For low numbered states the current vocabulary of physics is inappropriate . . . . For such states space and time do not exist – at least I can see no reason to believe that they do. But it must be supposed that when high numbered states are considered there will be found in the new scheme approximate counterparts of the space and time of current conception – something ready to merge into space and time when the state numbers are infinite.

About the old quantum theory, then, Eddington shows an excellent grasp of the achievements and, in addition, a good intuition of what would be needed next. I turn now to what did happen next. It became clear that the old quantum theory was in difficulties in a large number of places. It is certainly true that a few of these were human failings that could have been dealt with in time. For the most part they were more serious. Detailed troubles like that of the helium spectrum might, it seemed, be put right by greater knowledge of the structure of the helium atom. But no such remedy was possible for the fact that the rules of the theory referred only to periodic systems, such as an oscillating electron, although it seemed evident that Planck's constant must also have some part to play in the physics of non-periodic systems.

The new quantum theory

The log jam of ideas was broken in 1925 by Heisenberg. He set out from the critical position that the theory talked about entities like the position of an electron which were not observed. What was observed were the frequencies of the spectral lines. Heisenberg tried to construct a new theory entirely in terms of such observables. His early attempt to do this was followed in a couple of months by Born and Jordan who recognised that Heisenberg was in fact constructing afresh some mathematical entities – matrices – whose properties were well known. A matrix is a two-dimensional array of numbers, with rules allowing one to add and multiply the matrices (by combinations of sums and products of the constituent numbers). The numbers in the arrays were, more or less, Heisenberg's observable quantities; the whole array corresponded to the 'dynamical variables' like position, momentum, energy. It is not necessary to go into the details of matrix algebra. The point is, as Dirac saw in 1926, that the nature of quantum mechanics is captured by the way that the entities combine and the fact that this method of combination happens to be represented by the rules of matrix algebra is an irrelevancy. Dirac was later to express this point of view very clearly in his book (Dirac 1930). He describes a 'typical calculation' in quantum mechanics:

One is given that a system is in a certain state in which certain dynamical variables have certain values. This information is expressed by equations involving the symbols that denote the state and the dynamical variables. From these equations other equations are then deduced in accordance with the axioms governing the symbols and from the new equations physical conclusions are drawn. One does not anywhere specify the exact nature of the symbols employed, nor is such specification at all necessary. They are used all the time in an abstract way, the algebraic axioms that they satisfy and the connexion between equations involving them and physical conditions being all that is required. The axioms, together with this connexion, contain a number of physical laws, which cannot conveniently be analysed or even stated in any other way.

Dirac spoke of the observables or the dynamical variables as 'q-numbers', as distinct from the 'c-numbers' which were the quantities expressed by the usual numbers, or behaving just like them.

It will be noticed that Dirac sees here the necessity of reintroducing quantities, 'states' of a system, over and above the observables. Heisenberg's original critical programme proved too austere to carry out in full. What exactly constitutes a state is left unclear, though the act of 'preparing' a system in a certain state is clear; it is the result of performing certain observations on the system. That is, states are prepared by observables and observables act on states to produce new states. Part of the interpretation scheme for all this is the rule that if an observable O acts on a state s in such a way that

$$Os = ks,$$

where k is a number, then, in this very simple situation, k is the certain value of O when the observation is made on the state s. In more complex cases, when Os is not a multiple of s, a more complicated probabilistic interpretation is used. This does not concern me here.

Where had the classical mechanical picture ('a single electron orbiting a central proton' in the case of the hydrogen atom) gone to in this new formulation? The answer is that the classical picture is now a 'naming procedure'. One takes the classical expression for the total energy in terms of position variables q and momentum variables p. Then one says that the corresponding quantum system with the same name has the same expression for its total energy except that the qs and ps are now written in heavy type because they are q-numbers. This means that they satisfy a 'commutation rule'. If there is a single $\mathbf{q}$, corresponding to a particle moving along a straight line, with a corresponding single $\mathbf{p}$, the rule is

$$\mathbf{pq} - \mathbf{qp} = \hbar/\mathrm{i}.$$

Here i is the usual imaginary quantity of algebra, $\mathrm{i}^2 = -1$, and the way in which this plays an essential role is still a mystery. It was not immediately clear at the time how odd the occurrence of i was. The whole process I have just described is called 'quantisation'. The result of it is that the elements in the Heisenberg matrices are determined in such a way that the commutation rules are satisfied identically. The simplest non-trivial example is the harmonic oscillator, in which the energy is

$$E = \tfrac{1}{2}(\mathbf{p}^2/m + m\omega^2\mathbf{q}^2),$$

where m is the mass and ω the frequency. By an elegant argument which judiciously mixes physical interpretation and algebraic subtlety, it is then possible to show[4] that the condition

$$Es = es$$

is satisfied only if the number e is of the form $e = (n + \tfrac{1}{2})\hbar\omega$, for $n = 1,2,3\ldots$. This in itself was regarded as a success for the theory compared with the older one, for there half-integral quantum numbers were seen to be needed in some cases but there was no rationale for them.

Simultaneously, Schrödinger was publishing a series of six papers giving another version of quantisation which looked very different. He started from ideas of de Broglie but successively modified them until his version really amounted to giving another representation of the commutation rules. Taking again the case of one degree of freedom, Schrödinger's version was this: a state s is now represented by a function $s(q)$ of the classical variable q. The quantum variables, q, p (Dirac's q-numbers) are represented by *operators* acting on s; in fact,

$$\mathbf{q}s = qs, \quad \mathbf{p}s = \frac{\hbar}{i}\frac{ds}{dq}.$$

In these equations I have used s for the function which Schrödinger denoted by ψ and which is widely known as the 'psi-function'. My reason is that I am concerned to present quantum mechanics as Eddington saw it. It is true that he took over the ψ-notation but not the whole corpus of ideas that the notation suggests to modern eyes. The change of notation is meant as a reminder of this.

The substitution of p by $(\hbar/i)(d/dq)$ is one that succeeds in satisfying the commutation rule identically, because

$$\mathbf{p}\mathbf{q}s = \frac{\hbar d}{i dq}(qs) = \frac{\hbar}{i}s + \frac{\hbar}{i}q\frac{ds}{dq}.$$

Then the expression for the total energy of the system, when quantised, produces a differential operator on s and the resulting equation has to be solved subject to certain boundary conditions. Such a solution is possible only for certain values of the total energy. These values agreed exactly with those found by algebraic means for those problems that the algebraic technique had been able to master.[5] But Schrödinger's technique was able to solve many more problems because it recast the problem in a form which was much more congenial to theoretical physicists. They had at their disposal a wide range of techniques for solving rather similar equations describing water-waves, the vibration of membranes and strings and so on. Schrödinger was also able to show that, despite the different appearance, his theory was more or less the same as the Heisenberg–Born–Jordan–Dirac versions.

I have just described Schrödinger's earlier version of his theory, known now as the 'time-independent Schrödinger equation'. He went on from there to the 'time-dependent equation' in which the energy is no longer confined to a fixed value. In the equation for a free particle, i.e., under no forces, for instance,

$$E = p^2/2m,$$

the quantisation is carried out by replacing both E and p by differential operators, E being replaced by $-(\hbar/i)(\partial/\partial t)$. The negative sign arises from special relativity, and is of no great significance. In this more general version the quantised energy equation will contain the first time derivative of s. This time-dependent Schrödinger equation had many very successful applications in the succeeding years and it is only recently that questions have been raised about the internal consistency and coherence of the theory. Whatever the outcome of such questions, the equation does have one feature that played an important part in later thinking. Only the first time-derivative appears. This happens because the classical energy equation, which is to be quantised, is of the first power in E. The quantised equation is therefore an 'evolution equation'; given the state s of the system at one time, the equation shows how it evolves with time. A great deal of intellectual capital was invested in this aspect of the theory under the name of 'transformation theory'. It came to be recognised as essential that an equation to be quantised must be linear in the energy E.

The function introduced by Schrödinger is evidently of some physical significance but it turns out not to be directly measurable at an instant. One reason for this is that s has to be a complex number (not surprisingly since i is involved in Schrödinger's equation). Thus it has two degrees of freedom. If $s = t + iu$, where t and u are real, then the modulus of s, determined by $t^2 + u^2$ is measurable at an instant but the phase, determined by u/t is not. (It has recently been noticed that determination of phase is possible by means of an experiment lasting a long (theoretically infinite) time, but this does not invalidate the present discussion.) Thus s and $e^{if}s$, where f is real, represent the same state. This fact plays an important part in explaining how s is changed by a coordinate transformation, but the details are somewhat confusing (see Note 7 of Chapter 5).

I now turn to Eddington's reception of this new quantum theory of 1926 as he describes it in (Eddington 1928a). It is worth going into this in some detail because, with one exception to be explained later, this remained Eddington's notion of quantum mechanics. He took no part in the later mathematical developments of the theory but, from 1928 onwards, turned inwards to his own work. In the Gifford lectures he begins by describing three approaches which are 'almost three distinct theories', by which he means, respectively, the Heisenberg–Born–Jordan approach, Dirac's and Schrödinger's. Then he says:

For Dirac p is a symbol without any kind of numerical interpretation; he calls it a q-number, which is a way of saying that it is not a number at all.

I venture to think that there is an idea implied in Dirac's treatment which may have great philosophical significance, independently of any question of success in this particular application. The idea is that in digging deeper and deeper into that which lies at the base of physical phenomena we must be prepared to come to entities which, like many things in our conscious experience, are not measurable by numbers in any way;

He reinforces his view that this is the real lesson to be learnt from the developments of 1925–6:

If we are to discern controlling laws of Nature not dictated by the mind it would seem necessary to escape as far as possible from the cut-and-dried framework into which the mind is so ready to force everything that it experiences.

I think that in principle Dirac's method asserts this kind of emancipation.

Eddington found much less that was congenial in Schrödinger's mathematically convenient version:

Schrödinger's theory is now enjoying the full tide of popularity, partly because of intrinsic merit, but also, I suspect, partly because it is the only one of the three that is simple enough to be misunderstood.

and he registers the 'protest':

. . . whilst Schrödinger's theory is guiding us to sound and rapid progress in many of the mathematical problems confronting us and is indispensable in its practical utility, I do not see the least likelihood that his ideas will survive long in their present form.

One aspect of the new quantum theory which was seen by everyone to be of great importance arose in 1927, too late for Eddington to include it in his lectures. He added a note on it in the printed version:

It was Heisenberg again who set in motion the new development in the summer of 1927, and the consequences were further elucidated by Bohr. The outcome of it is a fundamental general principle which seems to rank in importance with the principle of relativity. I shall here call it the 'principle of indeterminacy'.

The gist of it can be stated as follows: *a particle may have position or it may have velocity but it cannot in any exact sense have both.*

In common with most physicists of the time, Eddington regarded this principle (now commonly called the 'uncertainty principle') as very fundamental:

The conditions of our exploration of the secrets of Nature are such that the more we bring to light the secret of position the more the secret of velocity is hidden . . ..
When we encounter unexpected obstacles in finding out something which we wish to know, there are two possible courses to take. It may be that the right course is to treat the obstacle as a spur to further efforts; but there is a second possibility – that we have been trying to find something which does not exist . . ..
When the concealment is found to be perfectly systematic, then we must banish the corresponding entity from the physical world. There is really no option.

He develops this theme a little later:

A New Epistemology. The principle of indeterminacy is epistemological. It reminds us once again that the world of physics is a world contemplated from within, surveyed by appliances which are part of it and subject to its laws. What the world might be deemed like if probed in some supernatural manner by applicances not furnished by itself we do not profess to know . . ..

The deliberate frustration of our efforts to bring knowledge of the microscopic world into orderly plan is a strong hint to alter the plan.

It means that we have been aiming at a false ideal of a complete description of the world. There has not yet been time to make serious search for a new epistemology adapted to these conditions. It has become doubtful whether it will ever be possible to construct a physical world solely out of the knowable – the guiding principle in our macroscopic theories.

Dirac's equation

By 1927 then, Eddington had absorbed the new quantum theory including the uncertainty principle, on which he put particular emphasis. But in 1928 there appeared two papers by Dirac (Dirac 1928a,b), written a month apart. Their appearance made for a major change in Eddington's ideas. The second paper does not concern me, as it simply gives a number of further developments and applications of the theory of the first one. In the first paper Dirac, who is often the most austere and mathematical of writers, gives a very full explanation of the context of his work. He begins by pointing to some of the remaining difficulties in quantum mechanics:

The new quantum mechanics, when applied to the problem of the structure of the atom with point-charge electrons, does not give results in agreement with experiment. The discrepancies consist of 'duplexity' phenomena, the observed number of stationary states for an electron in an atom being twice the number given by the theory.

He describes attempts to meet the difficulty and then continues:

The question remains as to why Nature should have chosen this particular model for the electron instead of being satisfied with the point-charge. One would like to find some incompleteness in the previous methods of applying quantum mechanics to the point-charge electron such that, when removed, the whole of the duplexity phenomena follow without arbitrary assumptions. In the present paper it is shown that this is the case, the incompleteness of the previous theories lying in their disagreement with relativity, or, alternatively, with the general transformation theory of quantum mechanics.

It is necessary to look in a little more detail at the mathematics involved here. The important part of Dirac's argument avoids complexity by looking at the free particle. The earlier expression $E = p^2/2m$ for its energy was the correct one in Newtonian mechanics and it was the quantisation of this equation that resulted in the new quantum theory. This expression has to be modified when special relativity is taken into account. This arose, as explained in Chapter 2, by seeking a transformation between inertial frames which leaves Maxwell's equations invariant. The result was that the Lorentz transformation:

$$x \to x' = \beta(x - vt),\ t \to t' = \beta(t - vx/c^2),$$

where $\beta = (1 - v^2/c^2)^{-\frac{1}{2}}$, has to be used in place of the so-called Galilean transformation:

$$x \to x' = x - vt,\ t \to t' = t.$$

These formulae are for the one-dimensional case and for simplicity I shall confine attention to this case, ignoring the y,z coordinates. Now in Chapter 5 I explained how physical quantities had to transform under a representation of the group. This has consequences for the position x and the momentum p of the classical system which, as we saw, acts as a naming procedure for the quantum one. In the first place, and not at all surprisingly, the coordinate x is not by itself a physical quantity, but the pair (x,t) is. A particle is not a geometrical point; it is at a certain point at a certain time. Turning now to the momentum, the mechanics of the situation makes it reasonably certain that it will be a physical quantity of the same general kind as the position and so it must also be part of a pair (p,W). Here W is, at present, some number which is needed so that the transformation of (p,W) will be of the same form as that of (x,t). This W still needs to be identified physically.

The interpretation of W can be carried out by looking at the special case of a particle at rest. For a particle moving with any speed, there is a reference frame in which, at any particular instant, it is at rest. For if the speed is v (which is any speed) the Lorentz transformation above, with the same v, will be that taking the frame to this 'rest-frame' of the particle. In this rest-frame, then, $v' = 0$, and so the momentum also vanishes, $p' = 0$. But W was introduced so that the transformation of p might be written

$$p \to p' = \beta(p - vW),\ W \to W' = \beta(W - vp/c^2).$$

Since $p' = 0$, the quantities in the frame in which the particle was moving must have been connected so that $p = vW$. Newton defined momentum p as mv, where m is the mass. This suggests interpreting W as the mass, but a little more subtlety is needed. Putting the value vW for p in the second transformation equation gives

$$W' = \beta(W - Wv^2/c^2) = W\sqrt{(1 - v^2/c^2)}.$$

But W' is the value of W measured in the rest-frame of the particle, and so it cannot depend on the motion of the particle but only on its intrinsic nature. Other values of W will differ from that one by a factor which depends on the velocity. Given the 'mass-like' nature of W this makes it clear that W', the W in the rest-frame, must be interpreted as the Newtonian mass of the particle, m. If W is called the mass, when the particle is moving with speed v, then we must use a different symbol than m for it, say M, and we must say that the mass M increases with velocity

$$M = \frac{m}{\sqrt{(1 - v^2/c^2)}}.$$

There is one more step in the argument before I reach the starting-point that Dirac took for granted. To say that the mass increases with velocity calls for some physical explanation of where the increase comes from. If the velocity is small compared with that of light, the expression for M is approximately $M = m + \frac{1}{2}mv^2/c^2$. For speeds that are not too large, then, the increase in mass is equal to the Newtonian kinetic energy, $\frac{1}{2}mv^2$, except for the factor c^2. This had led Einstein in 1905 to identify mass and energy, except for the factor c^2 introduced by historically based choice of units; that is, to put $E = Mc^2$. Since $P = Mv$, the corresponding expression for the energy in terms of the unincreased mass, m, and the momentum p, is

$$E^2 = p^2c^2 + m^2c^4.$$

This expression is the starting-point of Dirac's argument. He begins by referring to Gordon's suggestion in 1926 that one should start from this and proceed just as Schrödinger had done in the non-relativistic case. But, Dirac argues, this will not do. The resultant theory becomes such that it can answer only significantly fewer questions than in the Schrödinger case and this is because the equation is quadratic in the energy E and not linear. (The resultant equation is not an 'evolution equation'.) But because of the transformation of p and $M = W = E/c^2$ being linked together, linearity in E must be accompanied by linearity in p as well; so it is no use writing

$$E = \sqrt{(p^2c^2 + m^2c^4)}.$$

Instead Dirac introduces new 'dynamical variables', α, β for which

$$E = \alpha pc + \beta mc^2.$$

His notation is slightly different, and since he is considering the three-dimensional case, he had three αs instead of one, but I neglect that complication here. Since

$$E - \alpha pc - \beta mc^2 = 0,$$

it follows that

$$(E - \alpha pc - \beta mc^2)(E + \alpha pc + \beta mc^2) = 0$$

as well. Dirac chooses the new dynamical variables in such a way that this multiplies out to the correct energy equation. This will be so only if the variables satisfy

$$\alpha^2 = 1,\ \beta^2 = 1,\ \alpha\beta + \beta\alpha = 0.$$

The last of these equations is the surprising one, but it was less of a surprise to workers in quantum theory like Dirac who were accustomed to using q-numbers p,q which were such that $pq \neq qp$.

Dirac is then able to go on to show, by a painstaking construction, that the equation he has written, and the corresponding differential equation, are indeed Lorentz invariant. The quantities α,β can be represented by matrices with four rows and columns. But I am not concerned here with the mathematical complications. Eddington read Dirac's paper and saw that, at least in quantum mechanics, the tensor calculus as he had understood it up to then did not give all the invariant equations. He interpreted the difference between Dirac's work and his own as residing in the use of the algebraic symbols α and β, for which the multiplication rule shows that they cannot be numbers. He never notices the appearance of two-valued representations of the Lorentz group, though this does play an essential role in Dirac's treatment. Neither for that matter does Dirac. In some ways this omission is not serious, for the two aspects are complementary.[6] But the effect on Eddington was profound. He recalls it seven years later (Eddington 1936):

In 1928, P. A. M. Dirac made a bridge between quantum theory and relativity theory by his linear wave equation of the electron. This is the starting point of the development of relativity theory treated in this book.

Notice that at this stage Eddington regards his work as a further extension of relativity, a 'fourth step' after the three steps of special relativity, general relativity and Weyl's gauge theory. He continues to sum up the position in 1923:

Microscopic physics was the province of quantum theory; but in 1923 this was little more than a collection of empirical rules which led to no coherent outlook. The 'new quantum theory' began with Heisenberg's researches in 1925, and with the aid of many contributors it reached soon afterwards the current form generally called wave mechanics. The conditions were becoming ripe for a unification with macroscopic relativity theory.[7]

To say that Dirac's wave equation was the first connecting link gives only a partial idea of its importance. *It was a challenge to those who specialised in relativity theory . . .* We had claimed to have in the tensor calculus an ideal tool for dealing with all forms of invariance and covariance. But instead of using the orthodox tool Dirac proceeded by a way of his own, and produced an expression of very unsymmetrical appearance, which he showed to be invariant for the transformations of special relativity theory. Why had this type of invariance eluded the ordinary tensor calculus? As C. G. Darwin put it, 'it is rather disconcerting to find that apparently something has slipped through the net'. It was Darwin's insistence on this point in private conversation which led me to take up these investigations

I was soon convinced that this was the extension of relativity theory for which we had been waiting, and that Dirac's equation was only the beginning of a more far-reaching application of the methods and conceptions of relativity theory to microscopic phenomena. After seven years' work I find the possibilities latent in the new departure still far from exhausted.

To the details of that seven years work, and the succeeding ten years I turn in Part 2 of this book.

Notes

1. Rayleigh's 1900 formula derives from a straightforward equipartition of energy argument; it gives the energy density per unit change in frequency, e, as

$$e(f,T) = Cf^2T$$

for some constant C and this must be wrong because the total energy is then

$$\int_0^\infty e(f,T)df = \infty,$$

though it is a good approximation for low frequencies. Wien had already seen, in 1893–4 that

$$e(f,T) = f^3g(f/T)$$

('Wien's displacement law') by quite general thermodynamical arguments. Rayleigh's law comes by taking $g(x) = 1/x$. Careful measurements of e in Berlin led Wien to take

$$e(f,T) = Cf^3e^{-bf/T},$$

with a different constant C, which comes from the displacement law by taking $g(x) = e^{-bx}$. This agrees well for high frequencies and has the additional advantage that the total energy is

$$\int_0^\infty e(f,T)\mathrm{d}f = T^4 \int_0^\infty x^3 e^{-bx}\mathrm{d}x,$$

that is, proportional to T^4 as observed (Stefan's law).

2. Planck's calculation, which tacitly calls for Bose–Einstein statistics in the counting, gives as average energy

$$\frac{E}{e^{E/kT} - 1},$$

where k is Boltzmann's constant. But to agree with the displacement law, $E = hf$.

3. Confine attention to circular orbits. If the radius is a, then

$$\frac{e^2}{a^2} = \frac{mv^2}{a}$$

and the energy of the electron is

$$\tfrac{1}{2}mv^2 - e^2/a = -e^2/2a.$$

Now if $E = hf$, then h has the dimensions (energy) $\times$ (time) which reduces to the same dimensions as angular momentum. Bohr noticed that, if the nth allowed orbit was of radius a_n, velocity v_n and if its angular momentum $mv_n a_n$ was set equal to $nh/2\pi$, then eliminating v_n, a_n showed that

$$E_n = -\frac{2\pi^2 me^4}{h^2 n^2}.$$

The frequencies emitted will then be

$$f_{mn} = (E_m - E_n)/h = \frac{2\pi^2 me^4}{h^3}\left(\frac{1}{n^2} - \frac{1}{m^2}\right),$$

the formula known to spectroscopists for more than thirty years.

The fact that the constant $h/2\pi$ instead of h was used was explained on geometrical grounds. The condition $mv_n a_n = nh/2\pi$ is equivalent to $p_n c_n = nh$, where p_n is the momentum and c_n the circumference.

4. Dirac prefers to work with new variables

$$x = p(m\hbar\omega)^{-\frac{1}{2}},\ y = q(m\omega/\hbar)^{\frac{1}{2}}$$

for which

$$x^2 + y^2 = 2E/\hbar\omega,\ yx - xy = (qp - pq))/\hbar = \mathrm{i}.$$

Consider the observable $A = (x + \mathrm{i}y)(x - \mathrm{i}y)$ and consider states s for which

$$As = as.$$

Of course, $A = x^2 + y^2 + \mathrm{i}(yx - xy) = x^2 + y^2 - 1$ so the values a for A are closely related to those of E. For a real physical system it is clear that $x^2 + y^2$ cannot have any negative values. Also,

$$(x - iy)(x + iy) = x^2 + y^2 + 1 = A + 2.$$

Hence $(x - iy)A = (A + 2)(x - iy)$ and so

$$a(x - iy)s = (A + 2)(x - iy)s,$$

that is,

$$(A + 2)t = at, \text{ where } t = (x - iy)s.$$

So if a is a value of A, it is also a value of $A + 2$ and therefore $a - 2$ is another value of A. This argument fails if $a = 0$. The values of A must therefore be $0,2,4,6,\ldots$ (for the descending chain can only terminate at zero). Thus the values of E are $(n + \frac{1}{2})\hbar\omega$.

5. The harmonic oscillator in Schrödinger's treatment becomes the solution of

$$Es = -\frac{\hbar^2}{2m}\frac{d^2s}{dq^2} + \tfrac{1}{2}m\omega^2 q^2 s.$$

Using the same y is in Note 4, this becomes

$$\frac{d^2s}{dy^2} + (a - y^2)s = 0, \text{ where } a = 2E/\hbar\omega.$$

This is a well-known equation. The boundary condition that $s \to 0$ at infinity and is finite everywhere requires a to have the values $1,3,5,\ldots$ and the corresponding functions s are the Hermite functions

$$s_n(x) = e^{-\frac{1}{2}x^2}H_n(x),$$

where the H_n are the Hermite polynomials defined by

$$H_n(x) = (-1)^n e^{x^2}\left(\frac{d}{dx}\right)^n e^{-x^2}.$$

The details are not so important as the fact that the analysis was all available for other reasons.

6. In a more modern notation one may write Dirac's equation as

$$\gamma^a \partial_a s = ms$$

where

$$\gamma^a\gamma^b + \gamma^b\gamma^a = 2\eta^{ab}.$$

Performing a Lorentz transformation obviously leaves the equation unchanged if the γ^a are transformed like covariant vectors. But then a different set of γs arises. From the invariant form of the commutation relations, however, these new γs satisfy the same rules as the old, so the transformation of one set to the other is an automorphism of the algebra. Now the four γs generate, by multiplication, a basis for all such operators. This basis is

$$[1,\gamma^a,\gamma^a\gamma^b,\gamma^a\gamma^b\gamma^c,\gamma^1\gamma^2\gamma^3\gamma^4].$$

The algebra of such quantities is normal and simple; which implies that any automorphism is of the form

$$\gamma^a \rightarrow \gamma'^a = Q\gamma^a Q^{-1}.$$

If one then uses this to replace γ'^a in the transformed equation, one gets

$$Q\gamma^a \partial'_a Q^{-1} s = ms$$

and this has the same form as the original equation if

$$s' = Q^{-1}s,$$

which corresponds to a two-valued representation of the group (Q and $-Q$ give the same transformation).

Notice that one bonus from the new equation is a considerable simplification in the derivation of transformation laws when the coordinate-system is changed. The rule $s = Qs'$ corresponds here to $s' = e^{if}s$ of Note 7 of Chapter 5, but though the effect of the rule may be more complex (as would be expected since s now has four components) its status is that of a direct consequence of the algebra. On the other hand, this simple situation is restricted to the Lorentz group, whereas the earlier discussion of the Schrödinger theory applied more generally.

7. The situation was to prove more difficult than imagined. A very clear statement (Wigner 1957) has not since been bettered:

The difference . . . is, briefly, that while there are no conceptual problems to separate the theory of special relativity from quantum theory, there is hardly any common ground between the general theory of relativity and quantum mechanics. The statement, that there are no conceptual conflicts between quantum mechanics and the special theory, should not mean that the mathematical formulations of the two theories naturally mesh. This is not the case, and it required the very ingenious work of Tomonaga, Schwinger, Feynman, and Dyson to adjust quantum mechanics to the postulates of the special theory and this was so far successful only on the working level. What is meant is, rather, that the concepts which are used in quantum mechanics, measurements of positions, momenta, and the like, are the same concepts in terms of which the special relativistic postulate is formulated. Hence it is at least possible to formulate the requirement of special relativistic invariance for quantum theories and to ascertain whether these requirements are met. The fact that the answer is more nearly *no* than *yes*, that quantum mechanics has not yet been fully adjusted to the postulates of the special theory, is perhaps irritating.

PART 2

1928–33

7

Algebra to the fore: 136

Eddington's whole intellectual framework was shattered by Dirac's electron equation. The authoritative character of the pronouncements of MTR was hopelessly undermined by the evident falsehood of one of them: that all invariant equations were of tensor form. Relativity had by now become an accepted framework within which both Eddington and this young rising star (Dirac had come to St John's College as an unknown research student in 1923 and was 26 when he discovered the equation) were working. Yet somehow the content of relativity was different from Eddington's original conception.

The second part of this book traces the development of Eddington's ideas from the change of approach brought about by the advent of Dirac's equation to Eddington's attempt at an ordered presentation of his own theory in the first of the two books I have described as constituting the Eddington mystery, *Relativity Theory of Protons and Electrons* (Eddington 1936), referred to here as RTPE. In this chapter I shall be concerned with the crucial first stage in the development of the theory, for it was at that early stage that the possibility of the calculation of physical constants became apparent to Eddington. It was this that set him off along a path that his contemporaries could not or would not follow.

He came to see the algebraic structures arising from Dirac's original postulation as providing the clue to the union of relativity theory and quantum mechanics. Such a union was devoutly to be wished by many of Eddington's contemporaries in the 1930s. The difference between them and him arose because Eddington, on the analogy of Maxwell's unification of electricity and magnetism giving a value for the speed of light, expected and found values for other physical constants. I shall argue that initially these numbers arose for Eddington from a trial and error investigation of algebraic systems. As the investigation proceeded, Dirac's original equation, its later developments and Dirac himself moved into the background and the abstractions of algebra took over. Eddington's contemporaries gradually came to think that there was no more to it than that, that Eddington had convinced himself by manipulating the algebraic systems that certain accidental numbers were of great physical significance. From Eddington's philosophical point of view, as I shall show later in the chapter, this was a gross misjudgement; such calculations were perfectly plausible.

The eminent scientist

This change in Eddington's intellectual life was wholly independent of his public image. This had developed continually since the eclipse observations and continued to do so. His life in Cambridge was a busy one. In 1930 Einstein at last accepted a long-standing invitation to visit Eddington at the Observatory; in the same year Eddington was elected (in each case for two years) president of the Physical Society, the professional association of research physicists, and of the Mathematical Association, a society whose aim was to appeal to both university and school teachers of mathematics. It was also in this year that two notable honours were conferred on Eddington, his Knighthood and the Freedom of his home town, Kendal. In Cambridge the new statutes of 1926 had set up Faculty Boards; Professors were *ex officio* members and Eddington took his attendence at the board very seriously. He was also a member of the committee formulating the 'new' Tripos structure of 1935. His commitment to the Society of Friends led to his auditing the annual accounts of the Cambridge meeting. All this time he was still concerned with a modest amount of university teaching, with the conduct of the Observatory and with his own continuing researches in astrophysics. Between 1928 and 1933 twice as many of his twenty-five papers were on astrophysics as on his new theory. The controversies on the expanding universe (that is, whether the observed red shifts of the most distant objects should be interpreted as Doppler shifts caused by the expansion) around 1930 and on the age of the universe (stellar evidence seeming to require a much longer time-scale than the observations of distant galaxies) around 1935 found him fully involved.

Another less happy astrophysical controversy must be mentioned at greater length. It harks back to the internal constitution of the stars. Eddington had, as I explained in Chapter 2, modified the Hertz-sprung–Russell story of stellar evolution by requiring the perfect gas assumption to hold for the descending series of 'white dwarfs'. The results were in good accord with observation but there was a difficulty which I did not mention in Chapter 2. The white dwarfs were losing heat and so could be expected eventually to cool. This required them to revert to a lower density. Such a change could come, it seemed, only through expansion, a process requiring energy which the star did not have. This situation was saved (Fowler 1926) by the theory of a 'degenerate' gas. I will not pause here to discuss this notion of degeneracy in detail. Suffice it to say that the conditions of high density in white dwarfs means that the matter is in the form of an electron gas. The behaviour of an assembly of electrons, even under the perfect gas assumptions, will differ from that of an assembly of classical particles in virtue of the quantum mechanical properties of the electrons. It is not without interest that Dirac began in 1923 as a research student under Fowler, investigating the effects of relativity in statistical mechanics, but he soon moved over to quantum mechanics. Returning to the Fowler formula, this showed that cooling could occur with the density remaining high. Eddington found this explanation satisfactory. So matters rested for about three years, when it was suggested (Stoner 1929, 1930; Anderson 1929) that Fowler's formula needed modifying. The compression would, of course, have raised the electron speeds and these high speeds suggested the need for a relativistic correction to the formula, giving the 'relativistic degeneracy formula'. Chandrasekhar applied this formula to the white dwarfs (Chandrasekhar 1931) and so determined a limit to their possible masses of about one and one-half solar masses, the 'Chandrasekhar limit'. This restored the old difficulty that the original Fowler formula had banished. The star will have no equilibrium state, but will collapse continually to a 'black hole' from which light cannot escape.

Eddington could not bring himself to accept this; he attacked the Chandrasekhar limit and had a disagreeable quarrel with Chandrasekhar himself. He found himself forced to argue that 'there is no such thing as relativistic degeneracy'. The idea was based, he contended, on joining formulae from relativity mechanics and non-relativistic quantum mechanics. It was at this point that his astrophysical work first made contact with his new theory. By means of his new results he claimed to be able to show that a 'correct' allowance for relativity actually finished up with the original Fowler formula (a state of affairs which sounds very surprising but would not have been unique in relativistic calculations). Most astrophysicists were behind Chandrasekhar but Eddington was not to be budged. He was ready, in Bondi's opinion, 'to contort his physics' to retain the Fowler formula. He gave a new derivation of it in RTPE Chapter 13. It cannot be said to settle the matter; the argument extends over twenty-five pages but it is very obscure and to my mind not free of special pleading. Eddington's reputation was such that the Chandrasekhar limit was regarded with suspicion for something like twenty years. No one who knew Eddington was able to explain this myopic lapse from a scientific outlook though it must be noted that no one has in fact been able to say how a star of mass greater than the Chandrasekhar limit will evolve. It would be simplistic to make an obvious comparison and to see Eddington at forty-nine as a Master Builder still smarting under the shock of the young Dirac's discovery and determined to withstand the next assault. For one thing, the algebra had taken over and the personality of Dirac, if it were ever important, had ceased to be so. The most generous explanation for Eddington's uncharacteristic behaviour would perhaps be that he was genuinely misled by the necessarily incomplete form of his new theory. What is also remarkable is that Bohr, Pauli, Rosenfeld and Dirac separately told Chandrasekhar that he was correct but were unwilling to defend him publicly. This can be partly explained by their view that these astronomical matters were too recondite for physicists to pronounce on. But an important part of the explanation is Eddington's enormous reputation in the early 1930s. It was at its peak. Once RTPE had been published in 1936 the critics began to feel that something had gone wrong.

This busy Cambridge life was punctuated by a fair amount of foreign travel, usually to attend conferences or deliver public lectures. A public lecture that he gave at the meeting of the International Astronomical Union in the USA in 1932, together with the material of three talks broadcast there shortly afterwards, formed the basis for one of Eddington's most successful popular books (Eddington 1933a) *The Expanding Universe*. In 1934 he delivered the Messenger lectures at Cornell and these were published as *New Pathways in Science* (Eddington, 1935). His last trip abroad was in 1938, by which time the coming war was already casting its shadow.

Non-commutative algebra

It is hard to describe adequately the utter contrast between the outward manifestations of the great and respected scientist and the inward thoughts of the puzzled and dismayed thinker. In the gaps between his public life Eddington was privately engaged in trying to come to terms with the existence of Dirac's equation as described in Chapter 6. The key must lie, he felt, in the strange algebraic properties of the Dirac operators, and in particular, in the strange new observables α,β introduced by Dirac. Eddington's investigations began (Eddington 1928b) in a way which was, as he says himself, not very different from that of many others who had been attracted to the topic. But internal evidence suggests that it initiated a lengthy series of trial and error investigations of such algebraic structures. An intermediate stage in this series reached publication (Eddington 1931a, 1933b) but in a form very much influenced by the journal in which it appeared. (It was even more strongly the convention amongst mathematicians then than it remains now to write in such a way as to conceal all the effort that has gone into the derivations.) It is clear that this investigative process went on into the 1940s. It was protracted because Eddington liked 'to work these things out for myself'. This algebraic manipulation is a kind of ground base for the physical ideas, so something needs to be said about it. But we lack any papers Eddington may have left on it; though probably he would have consigned such experimental investigations to the waste paper basket.

I shall therefore tempt providence by attempting a bold reconstruction of his general early thinking about algebraic structures. It may be false to suggest that he was thinking on these lines, though I do not think it is. But some of the results of this thinking certainly form an important part of the starting point for his later work. To start this conjectural reconstruction, then, consider again Eddington's remark quoted at the end of Chapter 6 about Dirac's 'challenge to those who specialised in relativity theory'. He emphasised C. G. Darwin's 'it is rather disconcerting to find that apparently something has slipped through the net'. 'It was Darwin's insistence on this point in private conversation which led me to take up these investigations . . .'. We do not, of course, know the exact tenor of those private conversations nor indeed can we ever hope to know about many of the influences on Eddington. The strong collegiate nature of the university meant that at this time in his life, as distinct from later on, Eddington picked up information and suggestions informally, at High Table, attending mcetings of the Faculty Board or in a score of other ways. After 1936 Eddington turned in on himself, spent more time at the Observatory and so was less open to influence. However, we do have a letter that Darwin wrote as early as 11 January 1928 to Pauli. In it he says (Hermann, van Meyenn and Weisskopf 1979):

I do not know if you have heard that Dirac has got a new system of wave equations which does the whole spinning electron correctly, Thomas correction, relativity and all. He has told me something about them; they have four variables and are of the first order, but are *not* approximately equivalent to ours! I have not seen how they work out to give hydrogen, but they have one interesting thing, that is, that they are of course invariant for a change of axes, but not in vector form. I have amused myself by putting them into vector form, and the result is quite comical for it requires to express these four quantities no less than 2 invariants, 2 four-vectors and one six-vector . . . .. The whole theory of general relativity is based on the idea that form must be invariant for change of axes. Starting only from that principle, the theory develops the idea that everything must go into vector form – or tensor form, and the suggestion is that is the most general group of forms which can represent physical invariance. Has there been a fallacy somewhere in the relativity theory, and is it a lucky accident that the more restricted group of tensors is adequate for it? Another possibility is that the new case . . . is only invariant for orthogonal rectangular transformation, and that when it is desired to mix electricity with gravitation the preposterous vector system I have described above will become necessary.

How does this 'preposterous vector system', which Darwin had already by mid-January 1928, arise?

To guess Eddington's thinking on this and related questions one must go back from Dirac's paper to the simplest case of an abstract algebraic structure. Suppose that one has two objects e, f which 'combine' by some operation, which it is convenient to call multiplication (by analogy with numbers) to give two further objects ef, fe which are different. Further suppose that these new objects, and so presumably e, f as well, can occur both by themselves and (recalling $\alpha\beta = -\beta\alpha$ in Dirac's paper) multiplied by some positive or negative number.[1] This is all very tentative; I am trying to guess the way that Eddington sought understanding by a trial and error investigation. My justification for believing that this was Eddington's path is to be found in the manipulations on the lower half of p. 47 of RTPE. But there it is in more polished form and so the tentative thinking is obscured. For the sake of clarity I shall refer to the collection e, f, ef, fe and any other objects got in the same way by multiplications as an *algebra* and e, f, ef etc. as its *elements*. This is not to imply that Eddington formulated his language in this way at this early stage.

The simplest situation that one can envisage is that the new elements ef, fe

are not entirely independent, but that $ef = xfe$ where x is a number which is a constant for the particular algebra supposed to be generated by e and f. The next step in Eddington's thought is a much more subtle one. In Dirac's original paper there were three observables α, one for each of the three spatial dimensions. In the way that they occurred in the paper this was no great mystery. However, I am beginning, in this tentative reconstruction, with an abstract algebra generated by two elements e, f and an unspecified combining operation. Starting like that, it is utterly mysterious what meaning to attach to the notion of certain elements being 'associated with' spatial directions. I shall say more of this later. For the present I note that Eddington supposed that, in *some* way, e, f were associated with directions in space. Then, given that no direction in space is preferred to any other, it will follow similarly that, taking e, f in opposite order, $fe = xef$. Comparing these results gives $ef = xfe = x^2ef$ so that $x^2 = 1$. Since it is assumed that ef and fe are different, $x = -1$ and so $ef = -fe$ as Dirac says for α and β.

Next consider the element $g = ef$ which has been introduced by the combining operation. Then $fe = -g$. It follows that $g^2 = efef = -e^2f^2$ by interchanging the two middle symbols and remembering that $fe = -ef$. But also $g^2 = fefe = -f^2e^2$ in the same way, so that, although e, f do not 'commute', that is, $ef \neq fe$, e^2, f^2 do commute. This piece of deduction assumes that the so-far unspecified multiplication operation obeys the *associative law*, that $a(bc) = (ab)c$ for any elements a, b, c. Eddington would automatically have taken the multiplication to be associative without comment. Even amongst pure mathematicians there was virtually no interest in non-associative algebras, with the exception of one isolated system – 'Cayley numbers' – for another twenty years. Continuing the argument, by further uses of $ef = -fe$, $e^2f = -efe = fe^2$, $e^2g = e^2ef = -e^2fe = efe^2 = ge^2$ and in the same way $f^2e = ef^2$, $f^2g = gf^2$. These results suggest that e^2, f^2 must both belong to the set of elements commuting with all the elements of the algebra; at any rate, that would be the simplest arrangement. And the requirement of simplicity further suggests that this set of commuting elements should consist of multiples of one member, say u (for 'unit'), a unit element such that $ua = au = a$ for all elements a. If then $e^2 = pu$, $f^2 = qu$, where p, q are numbers, then it will follow that $g^2 = -pqu$. At this point, one can see that the absolute size of p, q is not very important. Thus, if $p = 4$, so that $e^2 = 4u$, it will simplify matters to choose a new symbol $E = \frac{1}{2}e$, instead of e, and then $E^2 = u$. But, of course, if p is negative this alteration of the structure by taking $E = p^{-\frac{1}{2}}e$ will not be possible. The best that can be managed is to finish with $E^2 = -u$ in that case. In all then one can choose $p,q = \pm 1$. It seems then that there are four possible combined values for p and q which we can list:

I $p = 1, q = 1$ so that $-pq = -1,$
II $p = 1, q = -1$ so that $-pq = 1,$
III $p = -1, q = 1$ so that $-pq = 1,$
IV $p = -1, q = -1$ so that $-pq = -1$

but the first three of these give essentially the same structure with two positive signs and one negative. Only the fourth one has the same sign for all so that the new element $g = ef$ is another element on exactly the same footing as e,f themselves. Once this situation has been reached a new possibility arises, bringing with it echoes from the mid-nineteenth century. If e,f are associated, in however mysterious a way, with two of the spatial directions, then g will naturally be associated with the third. Such an association is linked directly with the algebra of quaternions. In fact, the algebraic system IV of $3 + 1$ elements e,f,g,u whose squares are $-u, -u, -u,u$ respectively and such that

$$ef = -fe = g, \quad fg = -gf = e, \quad ge = -eg = f$$

is the *quaternion algebra*, Q, first found (in a different way) by Hamilton more than a century earlier (Hamilton 1844; Tait 1873). Hamilton wrote i,j,k where I have written e,f,g and he also made a simplification by identifying u with the number 1. The simplification is strictly illegitimate, but in fact does no harm and I shall follow it. The formal details are set out in Note 6 of Chapter 5, where it is made clear what constitutes 'association' of an algebraic element with a direction in space, for Hamilton. The apparent clarity of the explanation in the note is not to be found in Hamilton. It is just another example of the skill of the mathematician in concealing and by-passing difficulties. For it begins essentially with the geometrical directions and defines abstract algebraic relations between them. In the present tentative investigation the situation is the other way round; the abstract algebra is prior and the geometry is to come out of it in some as yet unexplained way. I think it likely that Eddington rediscovered Hamilton's algebra but did not realize until much later how it was related to Dirac's.[2]

Evidently this simplest algebra will not quite do for Dirac's equation, for it has only three anti-commuting elements instead of four but it teaches a useful lesson. One starts with a pair of anti-commuting elements whose squares are -1. The product of them is then a third one, in all respects similar. The complete set (including 1) has then three elements of one type (with squares − 1) and one of opposite type (square 1).[3] To anyone deeply immersed in relativity, where the first significant fact is that space-time, the arena for events, has three dimensions of one kind and one of another, this generation of 3 + 1 from algebraic considerations, mixed with a little symmetry, suggests at least a peculiar appropriateness for the algebra. And it pointed the way in Eddington's mind towards a very much bigger prize, for Eddington was, as I said above, an avowed Kantian. Now Kant argued that, firstly, space is not a conception derived from outward experience, since space is a prerequisite for such experience. Rather, space is a necessary representation *a priori*, the condition for the possibility of phenomena. It is important for Kant's general argument that geometry is a science which determines the properties of space synthetically and *a priori*. So the representation of space must be a pure intuition in the mind, prior to any perception of objects:

For geometrical principles are always apodeictic, that is, united with the consciousness of their necessity, as 'space has only three dimensions'. But propositions of this kind cannot be empirical judgments.

(Kant, 1787)

Eddington could not fail to be excited at the prospect of a new approach to complement Kant's discussion of three-dimensionality: if only one could somehow show the *unique* appropriateness of the algebra for the discussion of physical experiments one would have a derivation of the 3 + 1 dimensionality of space-time and so of the three-dimensionality of space. The virtue of this would be that Kant's admitted transcendental exposition would be replaced by one in much the same spirit but inside the theory. It was certainly worth trying, even if at this stage the task of carrying it through still seemed daunting. Eddington made a public conjecture about the interpretation of a closely corresponding result in the Dirac algebra (Eddington 1931b) but it is perhaps more significant that he even permits himself to speculate in the more austere pages of the London Mathematical Society (Eddington 1931a):

The partition . . . is probably of fundamental importance in physics, for I believe that the discrimination between time and space . . . is traceable to it.

But at that stage he still feels it necessary to qualify the statement:

Of course, it is not suggested that there can be any *a priori* proof that the fourth coordinate must be distinctive; but it is through this property of matrices that the distinction arises in a universe whose physical interactions are of a character postulated in modern wave mechanics.

Getting geometry from algebra

Eddington and a number of others began to interpret the Dirac algebra as related to space and time in somewhat the same way as the interpretation of quaternion algebra.[4] I shall describe this in a slightly modernised version. In this version the notation for the algebraic elements has been slightly changed so as to obviate the need for what Eddington calls 'reality conditions' which are a source of obscurity. In this more modern notation, Eddington, had he employed it, would have written four algebraic elements E_1, E_2, E_3, E_4 such that

$$E_1^2 = E_2^2 = E_3^2 = -1, \ E_4^2 = 1 \text{ and } E_a E_b = -E_b E_a \text{ when } a \neq b.$$

He would have called this 'the E-number algebra'. Taking account of the anti-commutation, there are then six products of pairs of Es, four of triples and one product of all four Es (corresponding to $g = ef$ in the quaternion case). Eddington used E_5 for this product, the square of which turns out to be -1.

It is well to notice here that the conventional approach, which Eddington shared to some extent, was to associate E_1, E_2, E_3 with three mutually perpendicular spatial directions and E_4 with time. So this association is quite different from the one adopted for quaternions, where the three spatial directions were associated with the two prior generating elements e, f *and* the derived one g. Here it is the four generating elements that are associated. The derived E_5 does play a physical role in the conventional theory but it is not that of an extra dimension. Eddington's thinking on E_5 was different. At various times he felt the need, either to have a five-dimensional theory to give E_5 an equal footing with the rest, or to identify E_4, E_5 with alternative time-scales in macroscopic and microscopic systems, or to go into more complex interpretations. Each of these attempts was in due course abandoned as unworkable.

When E_5 has been defined the sixteen elements generated by all the products of the various generators can conveniently be written as the $1 + 1 + 4 + 6 + 4$ elements:

$$1, \ E_5, E_a, E_a E_b, E_a E_5, \ (a,b = 1,2,3,4)$$

because, for instance, since $E_5 = E_1 E_2 E_3 E_4$, so $E_5 E_4 = -E_4 E_5 = E_1 E_2 E_3$, so that any triple products may be reduced to pairs. The products $E_a E_b, E_a E_5$ (when a,b are different) can be abbreviated to E_{ab}, E_{a5}, so that

$E_{ab} = - E_{ba}$ and so on. Since $E_{ab}^2 = E_{ab}E_{ab} = E_aE_bE_aE_b = - E_a^2E_b^2$, the elements of *this* type with negative square are those in which E_a^2, E_b^2 are both negative or both positive and so are three in number. In the same way three more elements E_{a5} with negative square arise, giving in all $1 + 3 + 3 + 3 = 10$. The sixteen elements are thus divided into ten of negative square and six of positive square. This division of 16 into $10 + 6$ would not have seemed so suggestive to Eddington as that of 4 into $3 + 1$, but he was not discouraged and, as will be seen below, further developments produced bigger surprises.

The next step in the interpretation is to look for the place where Lorentz transformations occur in the theory. In the quaternion case[4] the transformations of the form

$$u \to u' = quq^{-1},$$

where q is any element of the algebra, have a ready interpretation in terms of spatial rotations. Here, however, the same equation would have a q of sixteen possible elements. It is true that q may be multiplied through by any number, since q^{-1} will then be divided by the same number, but this still leaves q with fifteen degrees of freedom. The fifteen operations, just as with the quaternions, preserve sums and products and so preserve statements expressed in terms of them. But fifteen degrees of freedom was more than Eddington expected, for the analogy with quaternions would have suggested the Lorentz transformations and the rotations – six degrees of freedom in all. However, amongst all the possible qs there are six types of the form $c + dE_{ab}$ and Eddington readily noted that when $a,b = 1,2,3$ these produce spatial rotations about the axes of any 'vector'.

$$u = u_1E_1 + u_2E_2 + u_3E_3 + u_4E_4,$$

whilst if a or b is 4 the corresponding Lorentz transformations along the axes are produced. Eddington assumed that a more general q involving a sum of all six E_{ab} would give a general mixture of rotations and Lorentz transformations. In fact the situation is considerably more complicated (Hiley and Peat 1987) but the complications do not affect the use Eddington made of the result.[5] The physical significance of the full fifteen degrees of freedom will be discussed later (Chapter 9).

Since E_5 commutes with all the E_{ab} it is left unchanged by the six operations and so the terms attached to the E_{a5} in the general u are also components of vectors, like those attached to E_a.[6] The terms attached to E_{ab} in u are the six components of an anti-symmetric tensor of rank two (usually called a bivector). Here, then, are Darwin's '2 invariants, 2 four-vectors and one six-vector' (six-vector being a term used interchangeably with bivector). It is interesting to observe that Darwin had already seen the complexities that Dirac's algebra was introducing by January 1928, that is, a month or two after the discovery of the equation. Eddington's understanding progressed slowly and initially under Darwin's guidance. But by 1936 he had assimilated the ideas fully and he was really taking them more seriously than Darwin and elaborating their further consequences.

The 'double frame'

Eddington's next step was to ponder on Dirac's use of the E-number algebra to describe a single particle and to enquire what was to be done if not one but two particles were involved. This question arose for him for two quite separate reasons. The description of the hydrogen atom that Dirac had given was based on the notion of the nucleus as a fixed centre of force. The algebra arose only in the description of the relation of the electron to it. A full description, Eddington felt, would need to treat nucleus and electron on the same footing, and so deal with two particles. But another reason seemed to him more cogent. He felt that quantum mechanics erred in its frequent preoccupation with a single particle. This was, as he saw it, a neglect of the principle of relativity, that observed positions and velocities must be of one particle relative to others (so relative to at least one other). If this is taken into account – and many as well as Eddington were of the opinion that a union of quantum mechanics and special relativity was an important goal – there was again a need to describe a pair of particles, the object particle and the 'comparison particle', which Eddington saw as the idealised model of the environment assumed implicitly by quantum mechanics.

Now if Dirac were right, each of the particles would need elements of an E-number algebra to describe them. Could these be two different elements of the same algebra? In the case of the hydrogen atom the answer is none too clear; perhaps two elements of one E-algebra could serve to describe respectively nucleus and electron. But the need for a comparison particle is something different. Here the object particle and the comparison particle are quite independent and, Eddington felt, there was no alternative but to describe them by means of elements of two wholly independent algebras. To distinguish one from the other, Eddington used E for the elements of one and F for the other, though the Es and Fs taken by themselves are exactly similar to each other. Then, he argued, the combined system of object and comparison particles would be described by an algebraic system involving both, and so involving 'products' of Es and Fs. Since the algebras are wholly independent, this 'product' will not have any particular commutation structure, but instead $E_a F_b = F_b E_a$ and so on for all products of Es and Fs. Evidently this set of all products of the elements of the two algebras will have $16 \times 16 = 256$ degrees of freedom. By this physical argument Eddington rediscovered what the mathematicians call a direct product. To save introduction of new symbols like F_b the mathematicians prefer to introduce it in this way:

Given two algebraic structures A,B with m,n elements respectively, the direct product $A \times B$ is the structure whose elements are the mn pairs (a,b) of elements a,b one from each structure, with multiplication of such pairs defined by the rule that the first and second elements of pairs multiply independently of each other:

$$(a,b)(c,d) = (ac,bd).\text{[7]}$$

Amongst the 256 elements of the direct product of two E-number algebras those elements of positive square will arise from pairs of elements of the single algebra with the same squares as each other and so there are $10^2 + 6^2 = 136$ of these and the remaining 120 have negative square. The occurrence of the number 136 alerted Eddington to a further possible significance of the algebra, for it was by then a number of considerable interest in quantum theory. Indeed, the way in which the algebraic arguments had moved on from $4 = 3 + 1$ to $256 = 136 + 120$ gave him a profound conviction that he was on to something of tremendous importance, and so his efforts redoubled.

The fine-structure constant

To explain why this was so, I must go back a little to the old quantum theory. I mentioned in Chapter 6 that the Bohr model of the hydrogen atom was able to explain the sharp spectral lines very accurately by assuming that the electron moved round the proton in certain allowed orbits. The radius of the nth allowed orbit was $n^2 \hbar^2 / m e^2$ and the energy in that orbit turned out to be $- m e^4 / 2 n^2 \hbar^2$. (For the details see Note 3 of chapter 6.) In the early days of the old quantum theory this dynamical picture was taken quite seriously; there were not yet any questions about the true nature of electrons. The model was analogous to the corresponding gravitational problem of one planet in an orbit round the sun. But this analogy shows up the artificiality of the restriction to circular orbits; a more general model would have elliptical ones, just as in astronomy. A little more in the way of assumptions was needed to find the correct 'quantum conditions' for elliptical orbits,[8] but it was relatively straightforward to guess that there would now be two integer 'quantum numbers' say n and k. But when the energy is calculated the answer turns out to be $- me^4/2(n + k)^2\hbar^2$, that is, the same as in the previous case with $n + k$ written for n. Such a system is called 'degenerate'; any energy level, for instance $n = 3$ in the circular orbit formula, is now the same as $n = 2, k = 1; n = 1, k = 2; n = 3, k = 0$ of the elliptic case. 'Degeneracy' means that a number of states have the same energy level.

The situation changes when special relativity is taken into account, and it is necessary to do this because the electron speed in its orbit can be quite large. (It approaches one-hundredth of the speed of light in the lowest orbit.) Instead of the $n + k$ it now transpired (Sommerfeld 1916) that the expression for the energy involved $n + \sqrt{(k^2 - \alpha^2)}$, where α is the *fine-structure constant*, $e^2/\hbar c$, about 0.0073. The name arises because the formula gives, instead of a single spectral line as with the circular orbits, a close-knit group of lines, so exhibiting the 'fine-structure'. Thus the values of $n + \sqrt{(k^2 - \alpha^2)}$ in the example above would be 2.999973 and 2.999987 instead of 3. The degeneracy of the spectrum is removed and the fine-structure can actually be observed by more complex experiments. This is the historical reason for the name 'fine-structure constant' for $e^2/\hbar c$, but the number is then found to occur in a number of other contexts. That the name persisted emphasises again the obsessive preoccupation of the old quantum theory with spectroscopy. (The formula for the energy in circular orbits, for instance, can be written as $- \tfrac{1}{2}m\alpha^2/n^2$.) An important point is that, as can be seen from the appearance of α playing just the same role as the integer k under the square root sign, α is simply a number, independent of the units used in measurement.

Fairly early on in the discussion of this number, it came to be seen as arising, amongst other ways, as the ratio of two lengths that physicists had associated with the electron. One of these is the so-called Compton wavelength, $\hbar/mc$. Another is the 'classical electron radius', e^2/mc^2. The ratio of these, $\hbar c/e^2$, which is $1/\alpha$, is quite near to an integer in value. Originally the measurements suggested that this integer was 136 but later measurements favoured 137. More recent accurate measurements give

137.036, so that the hypothesis that α^{-1} is exactly an integer is ruled out. None the less, the notion of an 'essentially' integral value, masked perhaps by some experimental detail or other, needing a more complicated theory to unravel, persists. Thus F.J. Dyson (Dyson 1952) put forward an argument, in a rather different context, to show that $1/\alpha$ was the greatest number of electrons that could subsist in certain conditions according to quantum electrodynamics.[9] Confusingly, Eddington always uses the term 'fine-structure constant' for $1/\alpha$, rather than for α.

Calculating pure numbers

So much for the physical context in which Eddington came across a calculation which, beginning with purely algebraic considerations, found at the first stage 4 split into $3 + 1$ and at the third stage 256 split into $136 + 120$. This was the inspiration for the first of the papers which were to form the basis of RTPE (Eddington 1929) in which he argued that the value of the fine-structure constant was determined by the number of algebraic elements of a certain kind.[10] I shall not elaborate on his arguments in the 1929 paper here because they are in fact rather weak. Instead I shall deal in more detail in Chapter 8 with the greatly revised version of these arguments in RTPE.

We must not disregard the purely internal difficulties in making anything respectable of this tentative theory that algebraic structures might determine physical constants. These were, for instance, that 120 seems of no particular significance. At first, also, splitting 16 into $10 + 6$ was not particularly suggestive. And if the 'direct product trick' has been used twice to get significant numbers, should not further ones result from one, or two, further uses of it? Yet splitting at the next stage 65 536 into $32\,896 + 32\,640$ prompts no thoughts of physical interpretation. Indeed, this last was a serious difficulty that was to remain with Eddington – any algebraic machinery that he was able to construct, even if it gave physically interesting numbers, would always go on to give further ones of no conceivable interest. This offended against the eighth of Eddington's philosophical principles which I isolated in Chapter 4, that of non-redundancy. He was later forced by this circumstance to fall back on rather weak arguments to the effect that the simplest cases are of importance because physics does not seem to need to proceed to the more complex ones. But in 1928 and 1929 he seems not to have been particularly aware of this weakness in his position, and the other internal difficulties listed above could be brushed on one side to await the further elaboration of the theory.

But what of the external difficulties? These seemed much more serious. Once Eddington's paper had appeared his critics were more concerned to know why the number of elements in a certain structure should be in *any* way related to the ratio of hc to e^2, that is, to the ratio of two physically measured quantities which just happened to have the same dimensions of

$$(\text{mass}) \times (\text{length})^2/(\text{time})^2.$$

In Pauli's letters, for example, (Hermann *et al.* 1979) one finds Bohr writing to Pauli on 8 January 1929 'What do you make of Eddington's latest paper (136)?'. The editors of the Pauli volume rightly remark that this paper received wide notice because it raised the hope of determining the size of the electron charge by means of a union of quantum theory and relativity. And they add that the significance of the mysterious number 137 preoccupied even the sceptical Pauli in his later life: 'Perhaps it is not just an accident that the room of the clinic in Zürich where Pauli died was numbered 137' (my translation). There seems to have been no answer from Pauli to Bohr but by 16 May Pauli was writing to Sommerfeld 'I don't believe a syllable of Eddington's $\alpha = 1/136$.' And he had reached this conclusion earlier, for on 18 February he writes to O. Klein about the difficulties in quantum electrodynamics: "I fancy that only totally new ideas can get us forward. (Note, by the way: I now regard the Eddington '136-work' as complete nonsense; more precisely, as romantic poetry, not as physics.)" It is clear, then, at this stage that Eddington was tackling a widely recognised problem. There was natural scepticism at Eddington's highly doubtful solution, even though no-one else had any solution at all.

I mention these typical criticisms first because they show what Eddington realised he must try to answer when he came to write up this and later papers in RTPE. I want to concentrate on the form of the later chapter rather than the initial paper because there is a significant difference between them. By 1936 there could be no shadow of doubt that 136 would have, by hook or crook, to become 137. If then 136 was a number of 'degrees of freedom', possibly in a rather generalised sense, then an additional degree of freedom had to be found. In the search for this, which is the mainspring of the chapter, Eddington found what he considered to be a much better justification for the 136 as well.

Why it should be possible

Before I deal in Chapter 8 with the details of Eddington's Chapter 15 in RTPE, it is important to see the situation from Eddington's point of view rather than that of his critics. Pauli and his correspondents shared a kind of 'epistemological positivism' which seemed to have been a route for success in understanding the path that led to the new quantum theory (and was closely linked, as Forman (1971) argues, to the intellectual climate of the Weimar republic). To them science was not finding the essence of things, but formulating schemes, 'theories', to relate the empirical material. This bears only a superficial and misleading resemblance to Eddington's operational-ism. For him the formal schemes were intimately related to the 'conditions of the world'. Initially Eddington's British colleagues, nurtured in the Cavendish brand of English empiricism, were more sympathetic to him. Their support fell away only when Eddington's successive explanations remained obscure. They were pushed into the position of believing that Eddington, obsessed with 136 or 137, was deceiving himself into thinking he had a demonstration when none was possible. Thus Dirac's opinion, told by C. J. Eliezer (Taylor 1987) was: " 'He first proved for 136 and when experiment raised to 137, he gave a proof of that!' He sounded sceptical".

The possibility of such a demonstration, however, looks quite different from the philosophical position already formulated or implied by Eddington in *MTR*, but disregarded by his colleagues. (Of course, to argue that such a demonstration is possible is not to argue that one has actually been provided; that can be ascertained only by going into the detail.) Taking the eight points of view that I isolated in Chapter 4, the one which is least relevant is the one publicly expounded by Eddington, that of selective subjectivism. The other seven are of varying degrees of importance. The notion of falsifiability suggests the need for a bold hypothesis that can be tested; Eddington's hypothesis that a number of algebraic degrees of freedom is related to the fine-structure of hydrogen is suitably bold but its testability is less clear. The most obvious falsification would be the lack of agreement with the empirical data. But matters are not so clear cut. The latest figure does give a difference of $1/\alpha$ from integral value but only of about one part in 4000. This suggests a small correction factor, arising perhaps from some unknown further refinement in the theory. Moreover, the non-redundancy helps here. The algebraic formulation gives rise, in a very natural way, to a set of 136 elements. It is to be expected that it will be in some way identifiable.

It is with the ideas of operationalism, descriptive tolerance and theory-languages that something important about Eddington's outlook becomes apparent in the passage cited in discussing operationalism:

But we must not suppose that a law obeyed by the physical quantity necessarily has its seat in the world condition which that quantity 'stands for'; its origin may be disclosed by unravelling the series of operations of which the physical quantity is the result.

So here perhaps the algebraic structure is doing the unravelling for us. Why should such an obliging situation arise in connexion with the fine-structure constant? The idea of descriptive tolerance suggests the answer: somehow or other two language-games are going on (electromagnetism and quantum mechanics) and finding the relation between them is a process determining a constant, just as that between electricity and magnetism determined the speed of light. And here Eddington must have tacitly formulated an extension of the 'large-numbers argument' quoted in the discussion of the notion of theory-languages. There he was at pains to explain the occurrence of a large pure number (about 10^{40}); here the extension is simply that the largeness of the pure number does not imply any special status. Numbers of order 100 also deserve explanation.

Seeing science as concerned with structure rather than substance fits in very well with this. For any description or investigation of structure is likely to be best carried out by the branch of mathematics (group theory) in which structure is paramount. (Actually an apparently more general algebraic structure has been used but I have pointed out[1] that what is at issue is really group theory.) Finally, the principle of identification provides a possible way forward for the investigation. The earlier quotations about identification were very much couched in terms of tensors, because it was identification of tensors which was under investigation. But even there, as quoted in Chapter 4, Eddington looks forward to a situation in which:

We shall have constructed out of the primitive relation-structure a world of entities which behave in the same way as the quantities recognised in physical experiments. Physical theory can scarcely go further than this.

Eddington's critics, then, saw a hopeless task of such magnitude that his apparent achievement in carrying it out must be the illusion of someone bewitched by finding 136. Eddington, on the other hand, fascinated though he was by the unforced appearance of 136 and indeed of $4 = 3 + 1$, saw merely a relatively straightforward investigation, philosophically viable, but involving certain technical difficulties that needed ironing out. With these technical details I shall be concerned in the next two chapters.

I think it is worth going into these technical details for two reasons: firstly, the two arguments analysed seem to me to provide the most viable possibilities for a complete rewriting which would produce a modified argument acceptable to the most conventional physicists. I have not succeeded in carrying out this rewriting, but it is in any case outside the scope of this book. The second reason is that any reader with some technical knowledge of physics will want to have an analysis of at least one Eddington argument so that he can see the issues at stake: obscurity, failure to make premises clear and so on.

However, the technical details presented in these two chapters are not absolutely essential for my investigation of how Eddington came to write his two final books. The reader who finds the arguments daunting may proceed directly to Chapter 10, where the questions raised in Chapter 1 are made more precise and the first stage in answering them is reached.

Notes

1. The mathematical reader deserves some clarification at this point. I have used the vague phrase 'algebraic structure' because the use that Eddington makes of these structures is, for the most part, not as linear algebras. The other binary operation, addition, plays only a formal role. So essentially it is with groups that he is concerned. The ground field enters only in the form of its units; if the field is R, the group consists of $C_2 \times A$, where C_2 is the cyclic group of order two, $[1, -1]$ and A is the algebraic structure under consideration. If as with Eddington the ground field is C, the group is $C_4 \times A$.

2. By 1936 he was aware of two simplifying notations for the Dirac algebra and these are described in RTPE Chapter 3. One of these is in terms of three commuting permutation operators:

$$S_\alpha(1234) = 2143, \ S_\beta(1234) = 3412, \ S_\gamma(1234) = 4321,$$

and three commuting generalised permutation operators:

$$D_\alpha(1234) = 12\overline{3}\overline{4}, \ D_\beta(1234) = 1\overline{2}3\overline{4}, \ D_\gamma(1234) = 1\overline{2}\overline{3}4.$$

Since $S_a D_a = D_a S_a$, but $S_a D_b = - D_b S_a$ if $a,b = \alpha,\beta,\gamma$ and $a \neq b$, it is easy to show that the Dirac algebra is generated from products of Ss and Ds. The Ss and Ds, with unity, are both isomorphic to the quadratic group, $C_2 \times C_2$ (Klein's *Viergruppe*) and this exposes the Dirac algebra as a kind of skew direct product $S \times *S$. This analysis was to be found in (Eddington 1928b) but he also attributes to du Val the information that a similar analysis arises in connection with Kummer's quartic surface.

The second notation is in terms of two mutually commuting triads; the structure of each triad is then that of quaternions or, as a physicist would say, the Pauli spin matrices. Eddington attributes this to Lemaître, but he shows no realisation that this is a direct product construction.

3. Even the other choice of signs, although it lacks the symmetry of quaternion algebra, still finishes with a system, Q^*, in which three of the squares have one sign and one the other, although $+$ and $-$ are interchanged.

4. In the simple quaternion case, as explained in detail in Note 6 of Chapter 5, the association of the three elements e,f,g with three perpendicular spatial directions leads to the corresponding association of a vector $v = xe + yf + zg$ with the displacement from O to (x,y,z), referred to three spatial directions as axes. Then operations of the form

$$v \to v' = qvq^{-1},$$

where q is any quarternion, say,

$$q = a + be + cf + dg,$$

are all the spatial rotations. It is clear that q has four degrees of freedom whereas the set of rotations is described by only three. But this is as it should be, for if q is multiplied by any number, then q^{-1} is divided by the same number and the operation is left unchanged. The importance of operations of this type, from an algebraic point of view, is that they leave both sums and products unchanged. As a result, any statements in the algebra expressed in terms of sums and products only are left unchanged by the operations. This algebraic fact exactly mirrors the geometrical one that relations between points and lines expressed in terms of distance and angles are left unchanged by rotations.

5. I follow part of my essay in the reference cited. The complication arises because, for example,

$$(\cos \theta + E_{12}\sin \theta)(\cosh u + E_{34}\sinh u),$$

which arises by combining a rotation about the z-axis with a Lorentz transformation along it, contains also a term in E_5. Eddington omitted to notice the need to include an E_5 term. However, the set $[1,E_5,E_{ab}]$ *does* generate a sub-algebra; but, of course, it now has seven parameters which is one too many. There must be a restriction for it to represent the Lorentz group. To see what this is consider first what I call E-numbers of bivector type:

$$q = s + tE_5 + \tfrac{1}{2}w^{ab}E_{ab}, \text{ with the usual summation convention.}$$

Following an analogy with quaternions, define a conjugate E-number by

$$\bar{q} = s + tE_5 - \tfrac{1}{2}w^{ab}E_{ab},$$

so that

$$q\bar{q} = s^2 - t^2 + I_1(w) + 2E_5[st - I_2(w)]$$
$$= A(q) + E_5B(q) \text{ say.}$$

Here I_1,I_2 are the usual two invariants of w defined by

$$I_1(w) = \tfrac{1}{2}w^{ab}w_{ab}, \quad I_2(w) = \tfrac{1}{2}\tilde{w}^{ab}w_{ab},$$

where $\tilde{w}^{ab}$ is the dual bivector, $\tilde{w}_{ab} = \tfrac{1}{2}\varepsilon_{abcd}w^{cd}$. It follows that, since E_5 commutes with q,

$$q^{-1} = \frac{\bar{q}}{A + E_5 B} = \frac{\bar{q}(A - E_5 B)}{A^2 + B^2}.$$

I shall prove that the necessary and sufficient condition for this to be a Lorentz transformation is that $B = 0$; this is the single condition needed to reduce the number of parameters from seven to six.

The easiest way to prove this result is to revert to quaternions, in order to handle the algebra. The results got here will be useful in Chapter 9 in our revised treatment of Eddington's phase space. The three elements E_{23}, E_{31}, E_{12} have negative squares and anti-commute so they form a quaternion set e_1, e_2, e_3 (say). Then

$$E_{14} = - E_{1234}E_{23} = - E_5 E_{23} \text{ and so on.}$$

Since E_5 commutes with e_1, e_2, e_3 and has negative square, we write i for it and treat it (with care) as the usual imaginary. Thus

$$E_{14}, E_{24}, E_{34} = - ie_1, - ie_2, - ie_3,$$

and so

$$q = s + it + (w^{23} - iw^{14})e_1 + (w^{31} - iw^{24})e_2 + (w^{12} - iw^{34})e_3$$

a complex quaternion. $\bar{q}$ is then the usual conjugate quaternion, so the notation fits. Next consider E_4, which commutes with e_1, e_2, e_3 but anti-commutes with i (hence the necessity for the care mentioned). If E_4 is written as k, then

$$E_1, E_2, E_3 = - ike_1, - ike_2, - ike_3, \quad E_{45} = ki$$

and

$$E_{15}, E_{25}, E_{35} = - ke_1, - ke_2, - ke_3.$$

Now it has long been known (at least since Silberstein's efforts in 1912) that complex quaternions can be used to represent Lorentz transformations. The representation is rightly out of fashion because the introduction of the imaginary can lead to confusion with its use elsewhere, but no harm can come in this particular use. One writes

$$r = ct + i(e_1 x + e_2 y + e_3 z)$$

for a position vector. Then $r\bar{r} = c^2 t^2 - x^2 - y^2 - z^2$, so the norm is the relevant invariant. Such a quaternion r satisfies $r^+ = r$, where $r^+ = \bar{r}^*$, r^* being the complex conjugate quaternion. The condition $r = r^+$ ensures that r is physical. Consider now the transformation

$$r \to r' = qrs,$$

where q, s are complex quaternions. If this is to be a Lorentz transformation, then

(i) $(r')^+ = r'$, so that $s^+ rq^+ = qrs$ for all r.

(ii) $|r'| = |r|$ and this condition can be taken as $|q| = 1, |s| = 1$, since there is obviously some latitude in the definitions of q,s.

Taking these results together, $s = q^+$ and q is of unit norm, so the Lorentz group is contained in

$$r \to r' = qrq^+, |q| = 1.$$

But now q depends on three complex, that is, six real parameters, so this is exactly the proper Lorentz group.

In the E-number identification there is a slight difference. If

$$u = E_1 x + E_2 y + E_3 z + E_4 ct,$$

the corresponding quaternion form is

$$u = k(ct + ie_1 x + ie_2 y + ie_3 z) = kr.$$

So $u' = quq^{-1}$ becomes $kr' = q(kr)q^{-1} = kq*rq^{-1} = ksrs^+$, where $s = q*$. The required condition is therefore simply that $|q| = 1$, that is,

$$A(q) = 1, \ B(q) = 0.$$

If the second of these is satisfied, the first becomes a matter of normalisation, and so the theorem is proved.

An important notion in Eddington's theory is that of 'phase space' which is related to the automorphism group we have been discussing. The present result implies the need for fairly major changes to Eddington's theory of phase space, which will be discussed in Chapter 9.

6. If one includes improper transformations, such as space reversals, E_5 is a pseudoscalar (changes sign) and E_{a5} a pseudovector. Eddington did not concern himself with improper transformations.

7. It was not till later that Eddington noticed that the E-number algebra itself was a direct product. In our modernised version, it is $Q \times Q*$ because the four elements

$$(e,e),(f,e),(g,e),(1,f),$$

where the e,f in $Q*$ are those two with positive squares, have squares $-1, -1, -1, 1$, anti-commute and are easily seen to generate all the elements of the algebra.

8. The clue to the extension was to note that the Bohr condition $mv_n a_n = nh$ was of the form

$$\oint p \, dq = nh,$$

where p is momentum and q a coordinate. Confining attention to a two-dimensional orbit with Hamiltonian

$$H = (p_r^2 + p_\phi^2/r^2)/2m - e^2/r = E,$$

the ϕ-equation of motion gives $p_\phi = mr^2\dot{\phi} = $ constant $= K$ say. Instead of the other equation of motion, use conservation of energy to get

$$p_r^2 = 2mE + 2me^2/r - K^2/r^2.$$

Here E is negative for closed orbits. The quantum conditions are then

$$\int p_\phi\, d\phi = 2\pi K = kh,$$

$$\int p_r\, dr = 2\int_{r_1}^{r_2} \sqrt{(2mE + 2me^2/r - K^2/r^2)}\,dr = nh,$$

where r_1, r_2 are the roots of $p_r = 0$. The integral is elementary (though not particularly easy) and the result is found to be

$$2\pi[me^2/\sqrt{(-2mE)} - |K|] = nh.$$

Hence

$$nh = -kh + 2\pi me^2/\sqrt{(-2mE)},$$

so that

$$E = -\frac{me^4}{2(n+k)^2\hbar^2}.$$

9. Dyson's argument was directed to proving that the perturbation series solutions in quantum electrodynamics could not converge and so, since they gave accurate numerical results, they must be asymptotic series. But stripped of refinements, his argument is simply that, if N electrons are confined within a sphere of diameter of a single Compton wavelength, $\hbar/mc$, then their total Coulomb energy is $Ne^2/(\hbar/2mc) = 2N\alpha mc^2$. If this is greater than or equal to double the rest energy of an electron, that is, $2mc^2$, then 'pair creation' is possible, and so N ceases to be well defined. This condition is $N > 1/\alpha$.

10. I have used the sign of the squares of elements as a guide in the text. In fact, Eddington proceeds differently on different occasions. His favourite distinction seems to be based on the following: consider the transformation of the generating elements:

$$E_a \rightarrow (E_a^2)E_a = E_a^3$$

which will obviously result in $E_5 \rightarrow -E_5$ as well. Since, if we define

$$E_{ab} \rightarrow E_a^3 E_b^3 = E_a^2 E_b^2 E_{ab},$$

this transformation produces a division of elements into those which change sign and those which are left unchanged. The same numbers will result. But also, since $E_{ab}^3 = E_a E_b E_a E_b E_a E_b = -E_a^2 E_b^2 E_{ab} = E_a^2 E_b^2 E_{ba}$, products are changed in order, so that this is an involution (or, anti-automorphism) of the algebra. But this, and other apparently different distinctions, all turn out to be the same.

8

Electric charge: 137

The task ahead

If one tries to view the problem situation faced by Eddington from his point of view, it breaks into three parts. The expectation was that these three would prove connected in some way when the problem was solved. The first part of the problem is to make some connexion between 136 as a number of algebraic elements and 1/136 as related to (though by 1936 Eddington was aware that it was not equal to) $e^2/\hbar c$. As I said in the last chapter, the fine-structure constant occurs in many different contexts. Ideally, if it is to be related to algebraic structures, it should be possible to use any of these contexts with the principle of identification to establish the relation. Eddington chose, naturally enough, the occurrence of α as a coefficient in Dirac's equation of the hydrogen atom. Hitherto I have been speaking only of Dirac's equation for the free electron; that is the simplest case. But, as will have been clear from Dirac's explanation quoted in Chapter 6, his work was inspired – like so much work in both the old and new quantum theories – by difficulties in explaining the hydrogen spectrum fully. The way in which the equation for the free electron was modified to take account of the electrostatic force between proton and electron was well known, if not at all understood, in the quantum wave mechanics of Schrödinger. Dirac took it over without question and Eddington followed Dirac. No-one has since questioned this method of introducing an electric force, so I shall do no more now than to remark that it is a little mysterious. The method is simply to add to the energy operator, $(\hbar/i)(\partial/\partial t)$, an extra term corresponding to the 'Coulomb' potential energy in the field, e^2/r, so as to have $[(\hbar/i)(\partial/\partial t) + (e^2/r)]s$. When allowance is made for factors c in the rest of the equation, the result is an extra term $(\alpha/r)s$. So the first part of Eddington's problem was to relate the coefficient α in this modification of the electron equation, intended to realise the electron in a bound state orbiting the proton, to a number of algebraic degrees of freedom.

The second part of his problem was that all one could hope for from some reasonable sort of argument based on the general hints in the last chapter was a determination of the coefficient $1/\alpha$ as 136. This number of degrees of freedom needed to be increased by one. In retrospect, this proved a help rather than a hindrance; it made it clear to Eddington that the whole physical calculation could not be reduced to a mere counting of algebraic elements. This would be bound to give 136. It might well be that such a count would be needed somewhere in the argument, but the final answer would need physical arguments too.

The third part of the problem proved to be the least tractable. Eddington made a number of different attempts on it but none with complete success. It is this: the extra term in the Dirac equation to take account of the fact that it is the hydrogen atom which is under consideration is a product, $1/137r$. Eddington's argument has somehow to finish up with both factors in the product, only the 137 coming from the algebraic argument. And the factor $1/r$ is doubly important. It provides the legitimacy of the argument as it were. It allows for the use of the principle of identification to contend that this term acts just like the Coulomb potential so that no more is required. But also it serves as a marker; if the argument is to finish with a modified Dirac equation, then one must look in the result for the coefficient of the term in $1/r$.

A programme of analysis

That is a rational description of Eddington's position. In this and later chapters I shall analyse in detail how he tackled these problems. They are important for their own sake but my main reason for this detailed analysis is that Eddington's treatment of them is typical of his last books. Of course, the problems are not tackled in the order in which I have set them out. As with any other complex scientific work, a number of different approaches come together at the end to solve several problems at once. Correspondingly, I have to leave the chronological path through Eddington's life that I have been following up to now. I shall instead make a detailed analysis of two of the four principal arguments constructed by Eddington in RTPE. It will be sufficient to make only a quick reference to the other two later. The two arguments chosen for detailed scrutiny are each typical of its kind. The first one which I deal with in this chapter, on the value of the fine-structure constant, needs, in my opinion, only minor changes to make it logically complete. The problem is rather one of obscurity and originality and the major work called for here is simply to understand the argument. But the physical results of this argument follow only by substituting certain numbers which in RTPE are consequences of the other of my chosen arguments, that on the mass-ratio of the proton to the electron. This is another type of Eddingtonian argument. The general ideas put forward by Eddington are cogent enough, but the details do not fit them well. The problem here is less one of understanding than of reconstructing a defective argument and I make some tentative efforts towards this in the following chapter.

Returning to the subject matter of the present chapter, it is important to bear in mind one aspect of Eddington's treatment which will become apparent. It is that the discussion is carried out through a number of astonishing insights, each based on a particular model of the situation. These models are not usually stated explicitly, nor is any detailed relation between them explained. It is this unacknowledged changing between models rather than mathematical details which gives these later arguments their obscurity.

Eddington's argument

Eddington's argument for the fine-structure constant is a complex one which rests on four notions, three of which are original to him. But for some reason that is not clear, he chooses to present these three notions as essentially on the same footing as the fourth, that is, already current in quantum theory, though to some extent unacknowledged. He is to be seen, so his message runs, as the Socrates-like figure who questions quantum mechanics (and to some extent physics in general) about what it really means, so as to clarify matters. I shall deal with the four notions in the following order, which differs from Eddington's:

1. Indistinguishability and exclusion (which is the notion which is really present in quantum theory).

2.   Interchange (which Eddington sometimes seems to confuse with exchange, in the sense of exchange energy, in quanum theory).
3.   Comparison.
4.   Recoil.

My choice of order is to make the argument more comprehensible.

INDISTINGUISHABILITY

Fermi–Dirac statistics

The notion of indistinguishability is present in quantum mechanics but its importance is often underestimated. It depends on the curious fact that two similar quantum particles – two electrons, for example – are 'exactly similar'. That is, their two masses are identical in a different sense from the masses of two cricket balls. In the latter case we are aware that, although the masses agree sufficiently closely to fulfil the rules of cricket, yet a sufficiently accurate measurement of masses will always detect a difference. Not so with electrons; and this is related to the fact that a quantum particle has only a finite set of attributes.

Eddington saw the importance of this and realised that it must therefore play some part in the formalism of the theory. He approached the matter in the present argument by way of the strongest empirical evidence for it, that is, by way of the modification of statistics that it involves. This 'new statistics' had been introduced into quantum mechanics in the early 1920s. By 1936 Eddington had fully accepted that quantum mechanics implied a probabilistic description of events, so a statistical discussion was a natural way in. The details of the theory might be in doubt, the way of carrying out the description of the physical world might need modifying, but the basic notion was clear; there had to be a wave-function, more or less like that in the Dirac theory. The general problem was how to modify this description so that its relation with the highly successful but quite different formalism of general relativity could be exhibited. One feature of the quantum description which had been known for a decade was the need in considering assemblages of particles to 'use different statistics'. By this is really meant, not that anything is awry with statistics as usually understood, but that electrons, atoms and so on are not individuals of the type assumed in the statistics text-books. The difference can be made clear by a simple example.

Suppose a (very small) assemblage of three particles in a situation where, in accordance with the general ideas of the quantum theory, there are just four possible energy states which they might occupy. Further suppose (as is always necessary in quantum mechanics) that there is a complete description of the experimental set-up. In this case, assume that the energy is the only quantity which can be measured. In 'ordinary' statistics the number of ways in which such an arrangement can be made can easily be calculated as follows: the first particle can have one of four energies, whichever it has, the second can also have one of four and likewise the third, so there are in all $4 \times 4 \times 4 = 64$ ways. The probability of any particular configuration, say of (1,2), (2,4), (3,4), where (p,n) means that particle p has energy of the nth energy state, is then 1/64. But now suppose that the three particles are copies of one type of quantum particle and so are indistinguishable. Then the configuration quoted above cannot be distinguished from (2,2), (1,4), (3,4) or from (3,2), (1,4), (2,4). The total number of arrangements must now be calculated by simply asking how many triples of energies can be written down from four. It might be that all the energies written down are the same, which could arise in four ways, or two are the same and one different (twelve ways), or all different (four ways), so twenty in all. In this case, then, the probability of pattern (2,4,4) of energies is 1/20, that is, greater than 3/64 as it would have been using conventional statistics. A comparison of more use physically is to give actual values to the energies in the four states. For the sake of the example, suppose that the numbers 1,2,3,4 actually denote the values of the energy in the respective states, in suitable units. The total energy for the configuration under consideration is then 10. This can arise also as $3 + 3 + 4$ but in no other way (for only three particles) so that the probability of this total energy is $2 \times (1/20) = 1/10$ and so is greater than the 3/32 given by conventional statistics.

Exclusion

Such a new way of counting was called Bose–Einstein statistics. But spectroscopy (driving quantum mechanics as usual) had taken matters one step further. In explaining the complex structure of atomic spectra, use had to be made of the *Pauli Verbot*, or *exclusion principle*, which required that no two electrons in the atom might have 'the same set of quantum numbers', that is, be in the same state. It became clear that the characteristics of particles satisfying the exclusion principle was that they had 'half-integral spin', that is, were represented by spinor wave-functions, those transforming under two-valued representations of the rotation group. This gave rise to a new way of counting, called Fermi–Dirac statistics. The configuration listed above could not occur at all with Fermi–Dirac statistics (had probability zero) since it had two particles in the same energy state. And so also a total energy of ten units is impossible. The only possible energy totals turn out to be six, seven, eight and nine and each has probability $\frac{1}{4}$. With Bose–Einstein statistics the four probabilities are all 3/20 and with conventional statistics 5/32 for six and nine and 3/8 for seven and eight.

Eddington understood very well the need for different statistics for quantum particles; indeed, he was in advance of some in seeing this clearly as a method of description of unusual entities rather than as a defect in statistics. His first model, then, in the detailed argument, is that of an assemblage of particles, described statistically because it is too complicated to deal with them individually. But Eddington differed from everyone else in using this model to study a relation between exclusion and the electric force between charges:

It cannot seriously be maintained that the Coulomb force, which prevents two slow moving electrons from approaching one another, is an altogether distinct phenomenon from the exclusion principle (contained in Fermi–Dirac statistics) which achieves the same result by forbidding them to occupy the same cell . . .. This separation . . . strongly resembles the separation of gravitation and inertia in Newtonian mechanics . . .. There is need for the same kind of unification of treatment.

It will be noticed that this paragraph is more tentative than many of Eddington's. He cannot yet state a clear equivalence between exclusion and Coulomb charge, for not all excluding particles are charged and not all charged particles (e.g. charged particles not of half-integral spin are known) exclude. But one might be able to cope with the first of these exceptions by replacing neutral particles by some sort of compounds, and so perhaps some other sort of dodge might serve for the other difficulty. Eddington's tactics here are to regard the indistinguishability as logically prior. The Fermi–Dirac statistics are to be derived from it by asking how indistinguishability affects the wave-function description or, to use terms which Eddington has defined earlier in the book, affects the wave-tensors. He promises that what will result from this investigation is an extra term in Dirac's equation which is the same as the 'empirical' term of Coulomb energy.

INTERCHANGE

Superposition

In order to derive more consequences from the assumption of indistinguishability than had been found already in orthodox quantum mechanics, Eddington has recourse to the first of his three original notions, that of interchange. The model of an assemblage of particles described statistically is now dropped and replaced by that of a pair of particles. In due course these particles are to be taken as a pair of indistinguishable ones and so the consequences of the two notions can be derived.

In Eddington's first step of the argument to put some flesh on the bare bones of his programme he makes a fundamental use of the technique in the quantum formalism known as superposition. Some treatments of quantum mechanics (for example, Dirac's famous text-book) regard superposition as *the* characteristic of quantum theory. Briefly, it states that, if a system can be in one of two states, s,t say, then it is also possible for it to be in a superposition of these, denoted by $ps + qt$, where p,q are numbers. The rough idea about the values of p,q is that if one of them is much larger than the other, then the superposition is 'more in one state than the other'. Now suppose that s,t are characterised by certain sharp values of the energy, f,g. If E is the energy operator, this will be expressed in the formalism by

$$Es = fs, Et = gt.$$

Since E is a linear operator (also an unavoidable assumption of quantum mechanics):

$$E(ps + qt) = pEs + qEt = pfs + qgt,$$

so that the superposition does *not* correspond to a sharp energy state. The theory overcomes this hiatus by a fresh assumption: that the superposition corresponds to a state in which one or other of the sharp values of the energy will be observed but with a probability depending on the values of p,q. In fact, the probabilities are in the ratio $p^2 : q^2$, or, if p,q are complex numbers as is usual in quantum mechanics, the ratio is $|p|^2/|q|^2$. The details do not matter so much at this stage; what is important for Eddington's argument is the way in which the formalism 'blurs the sharp edges'. Instead of the two states, s,t one has a continuum of states and certain probabilities associated with them. Eddington puts it that he is going to follow quantum mechanics by replacing a jump by continuous

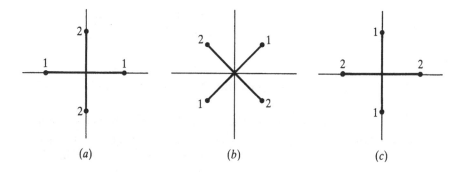

Fig. 8.1. Interchange as a continuous operation.

motion. The jump he has in mind is the interchange of two particles, and he is then going to ask for conditions under which this interchange makes no difference, because the particles are indistinguishable. He represents interchange by an operator Q in the form

$$Q(1,2) = (2,1).$$

For ordinary particles $Q^2 = 1$ but Eddington is now much clearer than he was in 1928 about the existence of two-valued representations, and he states that for spinors $Q^2 = -1$. This is just analogous to the way in which a rotation through four right-angles returns most objects to their former states, whereas spinors need eight right-angles because four just change the sign. Rotation is Eddington's guide here, though he puts the argument algebraically. One can see the result more clearly if one explores the geometrical picture behind his formalism. Fig. 8.1 shows a pair of rectangular axes; the one running from the first particle quoted (that is, in the pair (p,q) it represents p) and the vertical represents the other one. In the first figure (a) the heavy lines are labelled 1,2 respectively so the state $(1,2)$ is represented in (a) and $(2,1)$ in (c). The change from (a) to (c) is thought of by Eddington as a rotation, so an intermediate step is (b). Resolving along the fixed axes for the state, called it $(1,2,q)$:

$$(1,2,q) = [(1)\cos q/2 + (2)\sin q/2,(1)\sin q/2 + (2)\cos q/2]$$
$$= \cos q/2 \cdot (1,2) + \sin q/2 \cdot (2,1).$$

The lines are not to be thought to have a positive or negative direction associated with them, as they would if they were vectors. Then

$(1,2,0) = (1,2)$ and $(1,2,\pi) = (2,1)$ and the state $(1,2,q)$ is interpreted by Eddington as a superposition with probabilities $\cos^2 q/2, \sin^2 q/2$.

Indistinguishable particles

Now Eddington supposes the two particles to be indistinguishable so that the rotation has no effect on the system or in his language, it is a 'relativity transformation' to a new and equivalent reference frame, and in this situation he suddenly changes the guiding model. Although he expresses it in very general terms, it is clear that he is thinking of a dynamical model with which he is very familiar, that of a single planet orbiting the sun (the Kepler problem) (Fig. 8.2). Choosing as coordinates the radial distance r and an angular variable q, the expression for the total energy (the 'Hamiltonian' which is important in the quantum form of the problem) cannot contain q. This is because increasing q to $q + q_0$ merely corresponds to measuring q from a different initial line and does not change the system. On the other hand, both components of speed, along the direction of r increasing with fixed q and in the perpendicular direction produced by increasing q and keeping r fixed, will enter. In the terms usually used in mechanics, the momenta p_r, p_q will enter together with r. Then q is called an 'ignorable coordinate' though it should be emphasised that this is a technical term which is not to be interpreted in exactly the usual sense of the word. The analogy in the interchange situation is direct; the angle q is ignorable if the particles are indistinguishable so that it does not enter the total energy, but p_q does so. This result proves to be highly important, for p_q is to provide the extra term which gives the Coulomb energy.[1]

This section of the argument concludes with an unexpected gloss on the notion of indistinguishability, which has the effect of modifying the guiding model. This is presented in a characteristic Eddington way as a deepening of the concepts involved or, more correctly, as a pointing out that the concepts were in fact deeper than had originally been explained. Eddington is concerned to apply the ideas just formulated to the hydrogen atom, that is, to a system consisting of a central proton and a circulating electron. But whereas the electron has mass m and charge e, the proton has mass about $1836m$ and charge $- e$. Eddington's approach here is forthright: 'From the present point of view *protons are indistinguishable from electrons.*' The mass does not distinguish between them because 'mass can never be used as a

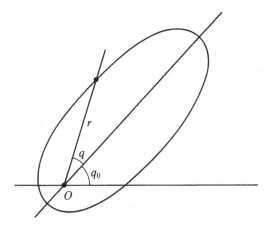

Fig. 8.2. The new guiding model: the Kepler problem.

criterion for distinguishing particles; it presupposes that they have already been distinguished.' The same goes for charge, but in any case there is no problem since the investigation is seeking to derive electric charge from prior considerations. I would not want to contest this insight of Eddington but only to point out that it is characteristic of his later arguments. The reader's interpretation of a relatively simple concept is suddenly found to be quite inadequate and has to be considerably deepened. Here he goes on to say that, contrary to the usual view, it is indistinguishable particles that are logically prior. 'The dynamics of distinguishable particles is a practical adaptation to be used when we do not wish to analyse the problem so deeply.'

Eddington's gloss on his statement is not a helpful one. A more illuminating view would be this one (which is modelled on some of his later thinking): a scientific theory is a developing entity, not given as a lump which is then gradually explained. It develops in the scientist's mind, partly as a result of thought, partly as a consequence of manipulations with bits of the physical world. It would be a gross error to use distinctions from a later developed part of the theory to criticise identifications in an earlier. For this point of view one can assert the indistinguishability of protons and electrons *at this stage*, since mass has not yet been introduced. If importing the temporal development of the theory seems to incongruous, one can rephrase the above argument in terms of theory-languages (see Chapter 4).

Conservation of probability

With one eye on the eventual outcome (136, 137 and the numbers of algebraic elements) Eddington now takes over the notion of interchange to E-number algebras. I omit the details of this because in fact the argument in the chapter I am considering never succeeds in making the necessary connections. Instead the numbers enter by a different route, using another model which I will describe later. It goes without saying that the algebras must play *some* part. That part will be made clear in the next section of my presentation of the argument, under Comparison. I proceed to the next step that Eddington takes, which involves another new model. The way in which Eddington uses this model makes it clear that he is at this point very much thinking in terms of the Schrödinger equation rather than the Dirac equation which has come to replace it and which is his real motivation. This does not matter because he is particularly concerned at this stage not with the details of the equation itself but with the interpretation (the 'Born interpretation') of the wave function s. The idea behind this interpretation is that the function s, which depends on time t and position P does not represent a physical field in the same way as the electric field E, which also depends on time and position and represents the electric force on a unit charge at P at time t. There is no suitable 'test-particle' to measure s.[2] But s is physically important and the proposal of the Born interpretation is that, if s is the wave-function of a single particle, then $|s|^2$ is a measure of the probability of finding the particle in a unit volume centred on P. There are two good reasons for this interpretation. Firstly, in the case of the hydrogen atom, which is the model for so much of the early development of quantum mechanics, the dependence of $|s|^2$ on r exhibits a number of maxima, of decreasing size as r increases, and these maxima are exactly at the radii of the Bohr orbits in the old quantum mechanics. So this interpretation gives a broad brush correspondence between the old and new theories. Secondly, if $|s|^2$ is such a probability, it must be something like the density of an imaginary fluid or gas – higher where the probability is greater, lower where it is less. But as the circumstances change and the probability is redistributed it will always remain the case that the particle will be found somewhere, so that the total probability added up over all the little volumes is one. That is, the imaginary gas may swirl about and be compressed or expanded but the total amount of gas is conserved. In the case of a real gas there is a well-known equation, that of continuity (see Note 1 of Chapter 4) which captures this.[3] The question arises, what is the corresponding condition on $|s|^2$ which ensures the conservation of probability? The answer is surprising – it is simply Schrödinger's equation again.[4]

The initial popularity of the Born interpretation was not based so much on these two arguments but on a misunderstanding amongst many physicists and one which persisted for a long time. It is the limitation in the above description to *s* being the wave-function for a single particle. For one particle there are three degrees of freedom in its position and so it is natural to see *s* as varying with position in physical space and to make the identification above of Bohr orbits in the physical space round the nucleus with the radii at which $|s|^2$ is maximum. But, though natural, it is not quite correct, as is shown up most clearly by considering the case of *two* particles. The number of degrees of freedom is now six and the *s* function must be interpreted as spread over them. A point in the six-dimensional space represents one particle being at one point in physical three-dimensional space and the other particle being at another. Thus the intuitive advantage of Schrödinger's wave mechanics over the Heisenberg formalism disappears for systems of more than one particle, as was already realised by Lorentz in a letter to Schrödinger on 27 May 1926. In it he explains how he sees wave mechanics as preferable for a single particle because of its intuitive clarity. 'With a greater number of degrees of freedom, however, I cannot interpret the waves and vibrations in *q*-space physically, and I must opt for matrix mechanics.' In any case, if the wave picture is pursued, then it is in the six-dimensional space that the probability is conserved.

To see the consequence of this, it is necessary to look a little more closely at the way in which the conservation of total probability – a global property of the system – is expressed as a local property, a differential equation of continuity, holding everywhere. It is easiest to see this by returning to the flow of an imaginary gas. Consider a small box with faces parallel to the coordinate-planes. Conservation implies that any change in the gas content of the box is produced by the flow in or out of it across the six faces. These six faces are in three parallel pairs. First consider one pair, say those parallel to the *yz*-plane. Only the *x*-component of velocity of the gas plays any part in transferring gas across this pair. Moreover, the net flow in will depend on the difference between the components of the velocity at the two faces. If the box is shrunk to a point, this difference is expressed by the *x*-derivative of the velocity. The same considerations apply to the other two pairs. So three spatial derivatives will arise in the conservation equation (as well as the time derivative which comes in because we are concerned with the change with time of the gas content of the box). In the six-dimensional case the argument is similar, so that the equation of continuity in differential form will there have six spatial terms. The same goes, of course, for Schrödinger's equation. Since this six-dimensional space cannot be physical space, it is clear that the three-dimensional one for a single particle cannot be either, for it is just the same kind of entity. Most of the time, in practice, no great harm comes from *identifying* it with physical space but it is necessary to remember that this is just a conventional identification, and one, moreover, which is possible only for the one-particle case.

I have expressed all the above argument in terms of Schrödinger's equation and the non-relativistic theory because that is the way that Eddington saw it. Naturally, somewhat similar considerations apply to the Dirac equation, though certain subtleties arise.[5]

The above paints the scene for Eddington's use of the idea of conserved probability. For two particles, he argues, when they are indistinguishable one must write, not (1,2) which would correspond to six degrees of freedom as above, but (1,2,q), with seven degrees of freedom. So the continuity equation and equivalently the Schrödinger equation will then have an extra term. This extra term, Eddington assures us, will turn out to be the required Coulomb field.

The hydrogen atom

The time has come for another shift of model, but this time Eddington gives an explicit signal. The hydrogen atom is his heading. This includes the Kepler problem of an earlier model but it includes the quantum theory of the hydrogen spectrum as well. The first step is to take the hydrogen atom as a two-particle system, the proton (which is the nucleus) of mass m_p, and the orbiting electron of mass m_e. Then this is further analysed by a trick familiar to astronomers into an 'external particle' of mass $M = m_p + m_e$ and an 'internal particle' of mass μ, where $\mu = m_p m_e/(m_p + m_e)$.[6] Eddington quotes some numerical values for these masses in terms of a 'standard mass' m_0:

$$M = \frac{136}{10}m_0, \mu = \frac{1}{136}m_0.$$

These so-called particles are not to be thought of, at least in astronomy, as having physical reality. The external particle is simply a conceptual entity exemplifying the motion of the centre of mass of the atom and the internal one similarly exemplifies the motion of the electron relative to the proton.

The notion of 'standard mass' will later need amplification. At this stage it is sufficient to think of it as merely providing a standard of mass. In the derivation of these equations, to be given in the next chapter, m_0 is determined as the mass of a 'comparison particle' whose existence is required according to the argument of this chapter under Comparison. The values given imply that m_p, m_e are the roots of a quadratic equation:

$$10x^2 - 136xm_0 + m_0^2 = 0.$$

The ratio of the roots of this equation is 1847.6, which is not the observed mass-ratio but is within 1% of it. These numerical values are quoted from an earlier chapter, Chapter 12, of RTPE. I shall not interrupt the flow of the argument by including the derivation of the numbers at this point. Instead, I shall simply assume them and explain what seems to me to be Eddington's most satisfactory derivation of the quadratic equation (he gave several) in the next chapter, as another example of his arguments.

The interchange momentum

Given the split between internal and external particles, Eddington now applies this to the notions of both time and interchange. Interchange does not, of course, affect the external particle at all, but with the passage of time the external particle may be affected by external forces in some way. The Schrödinger equation (for Eddington is still using the Schrödinger picture as his model) for the external particle has therefore four terms, three for the usual expression for the energy and one rate of change term.[7] For the internal particle Eddington insists that interchange must be allowed for, so that the internal energy has four terms, not three. But correspondingly there is no rate of change term here because the stationary states are sought, by using the time-independent Schrödinger equation for the internal particle.[8] In each case, then, the equation involves four terms and so the two equations have a superficial resemblance but this is misleading because the terms are not all the same.

Next, Eddington turns to the internal Hamiltonian (energy expression) H_i say. I said above that it has four terms rather than three. In finding the solution for the hydrogen atom, however, the four terms are not equally important. Classically the electron orbits are plane and the same is found to be true in wave mechanics. Ignoring interchange for a moment, the ordinary mechanical degrees of freedom correspond to motion in a plane and are conveniently described by plane polar coordinates (r,θ). The coordinate normal to the plane may be disregarded. The corresponding Schrödinger substitution for momentum is then fairly straightforward.[9] But now interchange needs to be taken into account as well. If proton and electron are interchanged, r becomes $-r$. By much the same argument as was used earlier to replace such a jump by a continuous process, Eddington proposes to replace the jump by an angular variable, ϕ say. Such an angular variable, he argues, will then give the same kind of term in the equation as that given by the other angular variable θ.[10]

It might be thought that the next step for Eddington would be simply to solve the Schrödinger equation with the interchange included and so to assess the consequences of including interchange. He does not do this, for good reason. It is by no means clear from the investigation so far exactly what form the Schrödinger equation will take. The reason for this lies with the coordinates used.[11] This is a straightforward difficulty in the process known as quantisation in quantum mechanics. The rule described above, of replacing the coordinates and momenta by operators, is not unambiguous because the operators may not commute. Thus, if x,p are the classical position and momentum, and if a term like x^2p^2 were to be present in the classical energy, the quantisation procedure as stated gives no preference between the different expressions $xpxp$, px^2p, x^2p^2 and so on. Coming back to Eddington's situation, the complete geometrical picture of r,θ,q or equivalently in Eddington's notation ϕ,θ,q is not available. Knowing only the conjugate momenta, the Schrödinger equation is not determined.

This is not an unusual situation in science. A direct onslaught on a position may be unavailable; in that case one usually uses what information there is, to find out what it is possible to do. Eddington is content to proceed in this way here.

COMPARISON

The material reference frame

At this point a third notion has to be brought in to bridge the gap. This notion arises as the most basic of Eddington's attempts to find common ground between relativity theory and quantum mechanics. It is set out fully at the beginning of a previous chapter of RTPE, Chapter 11. Naturally enough, a completely new model guides the development here. He begins by considering the effect of the *idea* of relativity on the classical notion of determining the position of a body. What is at issue here is more primitive than the formalism of either special or general relativity, though it underlies both. Galileo had already expressed it clearly in the seventeenth century. It is this: the position exists only relative to material objects (a fact particularly to the fore in an astronomer's thought). In the theoretical analysis, these 'reference objects' do not usually appear. Instead a 'geometrical frame' (coordinate-system) is a convenient surrogate.

The situation was not greatly changed in 1905 or 1915 by Einstein's two theories. But, argues Eddington, it *was* changed in 1926 by the new quantum theory, for that theory required objects to be known only probabilistically, and this requirement applies both to the object under consideration (fulfilling this forms the whole basis of wave mechanics) and to the reference objects. The sharply determined geometrical frame cannot then be a surrogate reference body, and so the material reference frame must now play some part in the theoretical description, although this is not realised in quantum mechanics. Naturally the environment tacitly assumed is a highly idealised and symmetric one. Eddington uses the term 'comparison fluid' for it. It is here that the need for a direct product construction, of the type which gave rise to 136, arises. If the object in question were described, by an imaginative extension of orthodox quantum mechanics, as it were, by an E-number, the comparison fluid will be described by something in the corresponding F-algebra, that is, in another algebra with just the same structure as the Es. The combined system of object + comparison fluid is then described by an element of the direct product EF-algebra of 256 degrees of freedom. The question that obviously arises is: how has orthodox quantum mechanics been able to be so successful without mentioning the comparison fluid? Eddington argues that this can come about in certain special circumstances, and the restriction that it should do so determines certain parameters. Instead of his detailed argument, I prefer a simplified example. Suppose that a system, ignoring its comparison fluid, is represented by an E-number P. As explained in the last chapter, the most general form of P is

$$P = A + BE_5 + C^a E_a + D^a E_{a5} + \tfrac{1}{2} G^{ab} E_{ab},$$

giving the '2 scalars, 2 vectors and a 6-vector' about which Darwin wrote to Pauli. The details of this form do not matter in this example. The most general comparison fluid, which is, by definition spatially isotropic, is of the form

$$C = H + KF_5 + LF_4 + MF_{45}.$$

The reason for this is that spatial isotropy prevents the spatial parts of the two vectors from appearing (for they would define a direction, in conflict with the notion of isotropy) and in the same way all of the bivector. Now the combined system consists of the direct product $P^* = PC$, an element of the algebra which has 19 of the possible 256 degrees of freedom.

Suppose that Q is another E-number, with a corresponding combined system $Q^* = QC$. The whole reason for representing objects by algebraic elements P,Q is that further algebraic operations may give new elements of importance. Now evidently

$$P^* + Q^* = PC + QC = (P + Q)C = (P + Q)^*.$$

That is, the addition operation functions in exactly the same way in the E-algebra (which corresponds in our example to the orthodox treatment, ignoring the environment) as in the EF-algebra. In this way, algebraic operations involving only addition can be carried out without error, ignoring the comparison fluid. The other important algebraic operation is multiplication. In that case

$$P^*Q^* = PCQC = PQC^2, \ (PQ)^* = PQC.$$

The multiplication will function in the same way in the E-algebra as in the EF-algebra only if $C^2 = C$. In the technical jargon, C must be idempotent. It turns out[12] that this is indeed possible for a C of the form given. In particular, H must have the value $\frac{1}{2}$. Eddington was aware of the existence of idempotent E-numbers with the scalar part having the value $\frac{1}{2}$ (though he attached more importance to those with scalar part $\frac{1}{4}$). He called them *compact* but said he was unaware of their physical significance. The present investigation makes this significance clear.

So much for the material reference frame. But a coordinate-system is also used in quantum theory; it can only be a calculational device if Eddington's argument is correct. Perhaps the distinction between a coordinate-system and a material reference-frame known only with uncertainty can be made negligibly small by using a very massive material frame? A massive system, rejoins Eddington, will cause a non-negligible curvature of space, according to general relativity, so that is no way out. On the other hand the suggestion does point to an important fact; the material environment tacitly assumed in quantum mechanics is a calculational tool closely related to the Riemann–Christoffel tensor in relativity.

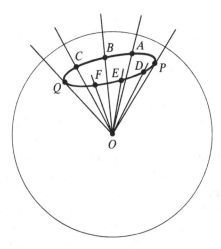

Fig. 8.3. Parallel displacement round a curve on a sphere.

Parallel displacement

How is this relation to be set up? In answering this Eddington is evidently influenced by the considerable success of quantum mechanics in discussing periodic systems like the hydrogen atom. Here the electron is in a stationary, roughly circular orbit. Bearing this geometrical picture in mind, he reverts to the notion of parallel displacement in general relativity and, in particular, to the change in a vector which is carried by parallel displacement round a small closed circuit. It is easiest to see what is at issue here by going back to the Gauss surface theory which I described in Chapter 3, in connexion with Einstein's discovery of relativity. There I noted that parallel displacement of a vector was constructed in two steps; a vector, initially in the tangent plane at a point P of the surface was displaced to a nearby point Q without any change at all. As a result of this first step the vector at Q will then have a component pointing out of the tangent plane because that tangent plane at Q will usually be different from that at P. The second step, which produces what is called parallel displacement in the surface geometry, is to ignore this normal component and to define the new vector at Q as having the components of the displaced vector in the tangent plane at Q. Consider for simplicity displacement of a vector round a small circle on the surface of a sphere (Fig. 8.3). The normal directions at the points of this small circle then all lie on a cone whose vertex is the centre of the sphere. As the vector is displaced from P, through $A, B, C \ldots$ to Q, it will be clear from the figure that the part of the vector along the normal will at each point have a component away from the reader and into the page. On the other hand, in going from P to Q along $D, E, F \ldots$ the corresponding normal parts to be thrown away are towards the reader and out of the page. The changes in the vector in parallel displacement from P to Q by the two different routes are therefore different. Let us put it this way: the vector v becomes respectively $v + v_1, v + v_2$. Consider now the effect of going from P to Q along the first path and then back to P along the other route; the vector will arrive at P again in the form $v + v_1 - v_2$ so that $v_1 - v_2$ will be the change in displacing the vector round a small closed circuit. It is evident that the size of this effect will depend on the radius of the sphere and on that of the small circle. The length of the path of displacement will be proportional to the radius r of the small circle. But as r increases so also does the angle of the cone and so the amount of change of the vector, so that the total effect (if r is small) is proportional to r^2. (Eddington puts it differently by relating it to the area πr^2 of the circle.) The radius of the sphere acts correspondingly in the opposite way, so the effect is proportional to r^2/R^2, where R is the radius of the sphere. Since $1/R^2$ is the sole component of the RC tensor for two-dimensional surfaces, one can see this as a product of the RC tensor and area.[13]

RECOIL

The comparison fluid

The calculation of parallel displacement which I have just described was a piece of mathematics well known to Eddington since 1923. It is in the new interpretation he gives it that the originality lies. He argues that when a vector suffers parallel displacement this is, as it is sometimes called, displacement without intrinsic change. The former analogy of surface geometry ('the part along the normal is to be ignored') is to be dropped, appropriately enough since the four-dimensional space-time is *not* given embedded in a higher-dimensional Euclidean world. How then can the change in the expression of the vector be interpreted, if there is no intrinsic change? Eddington's reply is that it represents the *recoil* of the comparison fluid. So here is another unsignalled change of model. The notion of recoil is very clear in classical mechanics but Eddington sees it as of more general applicability. In classical mechanics it is simply a consequence of conservation of momentum. It originally came to be noticed in gunnery. If a gun of mass M projects a shot of mass m with speed v by means of the (internal) explosion of some charge, the whole system of gun and shot cannot gain or lose momentum. Hence the gun recoils with a speed V such that $MV = mv$, so that $V = mv/M$. (In practice mechanical devices intervene to reduce the recoil as much and as quickly as possible.) Eddington now transposes the idea of recoil into a wholly new and more complex framework of ideas, a framework which had been set out, amongst much else, in his Chapter 11. A rather clear statement of the importance he attaches to it is to be found on p. 209 of RTPE:

The symmetry [of the comparison fluid] is disturbed by the recoil due to the motion of the object particle . . .. To neglect this recoil would be to neglect the very thing we are investigating, namely the mechanical specification of the object particle; for in the idealised universe consisting of an object particle and comparison fluid, the only manifestation of the mass of the object particle is in its mechanical reaction to the comparison fluid . . . the combination of action and reaction being expressed by double wave-functions [that is, in the *EF*-algebra]. But in applying this theory to the elementary wave-functions [for Eddington, in the *E*-algebra] of quantum theory, the complication arises that these are adapted to a different point of view. These simple wave-functions are intended to describe self-contained systems superposed on an *undisturbed* environment.

Returning to the particular case of the RC tensor, he says on p. 183 of RTPE:

The R–C tensor specifies directly the kinematical recoil of the comparison fluid corresponding to every possible cyclic displacement of the object system. If we introduce dynamical conceptions, we must attribute to the comparison fluid whatever dynamical characteristics are necessary to satisfy this specification, whether they are illustrated by dynamical models or not.

The last sentence is used to argue that mass and energy, represented in general relativity by the RC tensor, are inserted by this means into quantum mechanics. But it can also be seen as a first line of defence against possible criticism of the idea. This defence is carried further a little down the page:

From the standpoint of macroscopic theory the foregoing is a highly speculative interpretation of the R–C tensor. There is in fact an insuperable objection to adopting it in macroscopic relativity theory . . . . But in microscopic theory we approach it in a different way; the speculative taint is removed; and the aforementioned objection disappears as soon as we substitute displacement by wave propagation for the classical conception of displacement.

The objection mentioned is easily specified – the analysis assumes that the change in the vector arises uniformly round the cycle. But the removal of the 'speculative taint' is less straightforward. It hinges on the notion of comparison fluid and on the extent of the part played by convention in its definition. The observed displacements are relative to the comparison fluid but to use them in the theory, relative to a fixed geometrical frame, each observed displacement is analysed into a sum of two displacements, one of the object system and one of the comparison fluid, each relative to the reference frame. The partition of each observed value into two in this way is by convention. Different conventions correspond to different specifications of the comparison fluid.

How is this to be reconciled with the calculation above relating the RC tensor to the partition? Eddington replies that, if the geometrical frame has a metric, then its RC tensor determines the partition automatically:

... on the understanding that displacement without observable change is represented geometrically by parallel displacement . . . This double aspect [of the RC tensor] is the most essential link between relativity theory and wave mechanics.

The way in which Eddington now tries to put these general ideas into practice is somewhat tortuous. He contrasts two recoil situations. First, take the case when the internal particle has an observable displacement $d\theta^*$ (say) which – using the RC tensor partition just described to generate the masses μ, m_0 of internal particle and comparison fluid respectively – is analysed into a particle displacement $d\theta_p$ and a recoil $d\theta_c$ of the comparison fluid. Thus

$$d\theta^* = d\theta_p + d\theta_c, \ \mu d\theta_p = m_0 d\theta_c.$$

Solving,

$$d\theta_p = \frac{m_0}{m_0 + \mu} d\theta^* - \frac{136}{137} d\theta^*, \ d\theta_c - \frac{\mu}{m_0 + \mu} d\theta^* = \frac{1}{137} d\theta^*.$$

Second and in contrast the interchange dq is held to be conventional, not geometrical and so not based on the RC tensor. The most natural convention to take is to assume that the comparison fluid has no recoil, since it represents the array of macroscopic bodies whose individual recoils in quantum events will usually be negligible. How can this be incorporated?

Recoil of the comparison fluid

At this point Eddington reverts to the Dirac equation. It is probably not just a consequence of this change that he gives some evidence of intellectual discomfort. Four pages are occupied with somewhat muddled discussions of what would follow if this 'natural specification' were not the chosen one, so that the comparison fluid recoiled. I shall try in what follows to cut through the rather confused argument.[14]

Eddington's argument is that if only q could be treated just like θ the extra term resulting from interchange in the Dirac equation would be $(\hbar/ir)(\partial s/\partial q)$. This would result from analysing:

$$q = \frac{136}{137}q + \frac{1}{137}q,$$

the second term in the sum representing the recoil. If there is to be no recoil in q, this can be produced by a transformation of the geometrical frame of the comparison fluid by the same amount. At this point Eddington reverts to the argument by which he was initially led to a direct product construction. The comparison fluid is to be described by the other E-number algebra for which Eddington uses the F notation.

The calculation depends on the invariance properties of the Dirac equation (the very constituent of the surprise for Eddington and others in 1928). Dirac showed by a painstaking explicit calculation that his equation was Lorentz invariant. The form which his calculation took was to begin with the equation as originally written down and to change this by performing an explicit Lorentz transformation on the coordinates x,y,z,t involved. The new equation could then be shown to be brought back to the same form as the original if the anti-commuting elements α, β (or in Eddington's later notation, the E_a) were also changed by operations of the form

$$\alpha \rightarrow \alpha' = Q\alpha Q^{-1}.$$

Here Q is an element of the Dirac algebra or, in Eddington's later version, a general E-number. The transformation is just like that considered in Note 5 of Chapter 7. If Dirac's equation is written in abbreviated form as

$$Ps = ms,$$

where P stands for the momentum operator, including the necessary αs and

β or, in Eddington's notation, the *E*-numbers, and also the differential operators standing for momentum, the transformed version is

$$QP'Q^{-1}s = ms.$$

When this has been done there is still the state function *s* to be considered. Now if

$$s = Qs',$$

the equation becomes

$$QP'Q^{-1}\cdot Qs' = mQs',$$

that is,

$$P's' = ms'.$$

It is this that serves to show that the state function *s* transforms as a spinor, that is, under a two-valued representation. For if *Q* is replaced by $-Q$ the *transformation* $Q(\ )Q^{-1}$ is unchanged, whereas *s* is changed in sign.

The fine-structure constant

The situation in Eddington's argument can now be tackled. There is to be a coordinate transformation to remove the recoil of the comparison fluid which, though non-existent, had been allowed to enter the theory in order to derive the extra term corresponding to the momentum of interchange by comparing it with momentum for an angular variable. But there is one crucial difference. Ideally the whole discussion should have been carried out in the *EF*-algebra, so as to consider both object and comparison fluid, and it is in the comparison fluid where the recoil has to be removed. The original Dirac equation, written in the *E*-algebra, must be more or less equivalent to one in the *EF*-algebra. It is then in the *F*-algebra that the *Q* must operate, and in which *s* must be replaced by $s' = Q^{-1}s$.

At this point, in an aside, Eddington makes a curious error. He takes a Q^{-1} of the simple kind considered in Note 5 of Chapter 7 corresponding to a Lorentz transformation. Now in fact the interchange *q* has been taken as an angular variable, so that the recoil is of the same kind, a rotation, and the correct transformation will be

$$s' = (\cos q/2.137 - F_{23}\sin q/2.137)s.$$

Here F_{23} has been selected; this has the effect of partly specifying the coordinates for the comparison fluid so that the angle of recoil is in the *yz*-plane. The Dirac operator, which we wrote as *D* above, is not now Dirac's original *D* because it has the extra term $-(\hbar/r)(\partial/\partial q)$ corresponding to the interchange momentum. Even so, it is not affected by the transformation, since it lies in the *E*-algebra, not the *F*-algebra. The effect of the transformation is therefore to leave everything essentially unchanged except that $-(\hbar/r)(\partial s/\partial q)$ is replaced by

$$-\frac{\hbar}{r}\frac{\partial}{\partial q}\left[\left(\cos\frac{q}{2.137}-F_{23}\sin\frac{q}{2.137}\right)s\right]$$

$$=\left(\cos\frac{q}{2.137}-F_{23}\sin\frac{q}{2.137}\right)\left[-\hbar\left(\frac{1}{r}\frac{\partial s}{\partial q}-\frac{F_{23}}{2.137r}s\right)\right].$$

There is, therefore, an extra term $\hbar F_{23}/2.137r$ in the operator which acts on *s*, and this term results from the interchange momentum and the non-recoil of the comparison fluid.

This result resembles the Coulomb term which Eddington claims to be deriving but there are two differences:

(i)   There is a factor 2. The term has only half the expected value.
(ii) There is also a factor F_{23}, whereas the corresponding Coulomb term (taking account of the factors in the rest of the equation) would have had the algebraic imaginary i.

Eddington deals with these two separately. The factor 2 is to be expected, according to a much earlier discussion of his in his Chapter 9. In Section 9.6 he gives a rather confusing account which is, I believe, equivalent to the following: the correct state function *s* in the double-frame (*EF*-algebra) formulation is the product $s = s_0 s_c$ of the state-functions for object system and comparison fluid. Each factor acquires a transformation $Q^{-1} = \cos q/2.137 - F_{23}\sin q/2.137$. In the same way *D* is transformed by QDQ^{-1} but because *D* is an *E*-number and *Q* an *F*-number, $QDQ^{-1} = D$. As far as *D* is concerned *Q* has no effect. But this is not the case for s_0, since the state functions are not elements of the algebra. Thus the overall effect is

$$Q^{-2} = \cos\frac{q}{137} - F_{23}\sin\frac{q}{137}.$$

There is no doubt of the essential correctness of this.

It is quite otherwise with Eddington's solution of the problem (ii). On p. 298 of RTPE he says:

The transformation due to the rotation of the frame is therefore

$$\psi = e^{-\frac{1}{2}i\chi_0}\psi_r \ldots$$

The transformation matrix is properly the matrix F_{r4} associated with the ... rotation of the comparison fluid in its own frame F_μ; but since F_{r4} commutes with all the symbols, it is for our purposes an algebraic square root of -1.

That is to say, he simply replaces it by i so as to agree with the usual equation. For F_{r4} one should read F_{23} as explained above. That error is not serious. But I would argue that the remaining point is a major hiatus in the argument, perhaps the only one. Eddington's last sentence really will not do. It is doubtful in the first place how one could ever ignore the special algebraic qualities of F_{r4} (or F_{23}). But even if that is sometimes possible it will not suffice here, for to derive the correct result F_{r4} is not to be equated to '*an* algebraic square root of -1' but to the particular one, i, which occurs in the rest of the equation. (Even if an argument for this could be provided, it would inevitably leave a question of sign, $\pm$ i, undetermined.)

There is too much that is good, original and inspiring in Eddington's argument to reject it entirely for this one *non-sequitur*. But it obviously calls for further investigation. It would be beyond the limits of the present book to undertake this but I will point out what I believe to be the most promising approach. The appearance of the algebraic i is a source of great mystery in quantum mechanics; it was so to Schrödinger when he found himself compelled to introduce it. I think that this mystery and the difficulty (ii) may turn out to be related, and so both solved together. By this I mean that the intrusive i in quantum mechanics arises because the comparison fluid has been ignored there. If it were taken into account from the beginning, then it would need an F-element to do it. I believe that this element is replaced by the factor i in the usual formulation, which shows in this way that it is not quite possible to leave the environment out of consideration after all.

A summary of the argument

This chapter has inevitably been a complicated one. It has dealt in detail with what I regard as the most defensible of Eddington's arguments in RTPE and so I have been at pains to describe it at length. It seems useful to conclude this chapter by rewriting it in summary form. This could not have been done earlier, because it was necessary to explain the new concepts that the argument introduces.

The first of these new concepts is that of *indistinguishability*, used in a more general sense than usual (since proton and electron may be indistinguishable in suitable circumstances). This is based, like the corresponding notion in orthodox theory, on Fermi–Dirac statistics for an ensemble; and this also relies on *exclusion*. It is promised that the argument will show that exclusion gives the Coulomb term e^2/r. But that is in an ensemble; the next question is how to introduce exclusion and Fermi–Dirac statistics into a two-particle system. This introduces the new concept of interchange which, following the usual pattern in quantum mechanics, is smoothed out – to a rotation in an imaginary space, so that there is an interchange coordinate q. If the particles are indistinguishable this coordinate is a 'relativity rotation' so, on the analogue of the Kepler problem, q is not in the Hamiltonian though p_q can be. Carrying this over to the notion of probability and its conservation, which is embodied in the Schrödinger equation (or, though with difficulties, the Dirac equation) the extra probability in the q-direction will give rise to an extra term in the equation for the hydrogen atom.

This atom is split by means of the two-particle transformation of classical mechanics into an external particle of mass M and an internal one of mass μ. Values for these are quoted from another argument to be described in the next chapter of this book. In the internal case the specific angular character of the coordinate q comes by noting that, since interchange is changing r to $-r$, it is useful to put $\phi = \log r$, so that change of sign of r is produced by increasing ϕ by $i\pi$. The momentum is

$$\frac{\hbar}{i}\frac{\partial}{\partial(iq)} = -\hbar\frac{\partial}{\partial q}.$$

Since common ground is being sought between general relativity and quantum mechanics the next concept to be introduced is that of comparison. The argument is that there is need for a reference body, which in quantum mechanics is the 'comparison fluid' and so for the EF-algebra. This fluid, for periodic systems, recoils and the RC tensor explains how in terms of displacement without intrinsic change. But should the comparison fluid recoil for interchange? This is shown to be a question of convention, and the natural convention is to say that it does not recoil. The easiest way to incorporate this into the mathematics is to treat q at first like other angular variables and then to use a coordinate transformation to remove the recoil again. This readily gives the extra term for the Coulomb energy but with one real difficulty. This difficulty, that F_{23} is written instead of i, can perhaps be solved only by recognising that the intrusive i in quantum mechanics comes from the need to consider the environment (comparison fluid).

Notes

1. The inclusion of interchange (by Eddington's method) in the classical Kepler problem simply leads to a rotating orbit: the classical Hamiltonian is

$$H = \frac{1}{2\mu}\left(p_r^2 + \frac{1}{r^2}p_\theta^2\right) - \frac{G}{r}.$$

Eddington's modification of putting $r = \log\phi$ will then give rise to a Hamiltonian of the form

$$H = \frac{1}{2\mu}\left[p_r^2 + \frac{1}{r^2}(p_\theta^2 + p_q^2)\right] - \frac{G}{r}$$

and the equations of motion

$$\dot{p}_\theta = 0, \ \dot{p}_q = 0.$$

The first integrals $p_\theta = H$, $p_q = K$ are easily shown to lead to the energy equation:

$$\tfrac{1}{2}\mu\dot{r}^2 = E + \frac{G}{r} - \frac{1}{2\mu}\frac{K^2 + H^2}{r^2},$$

and this translates into the usual u,θ notation as

$$\frac{d^2u}{d\theta^2} + \alpha^2 u = \frac{\mu G}{H^2},$$

where $\alpha^2 = 1 + K^2/H^2$, giving the usual result for a rotating orbit. On the other hand, if this system is quantised in the manner of the old quantum theory by putting $H = kh$ and $K = nh$, whilst

$$2\pi p\hbar = 2\int_{r_1}^{r_2}\sqrt{\left(2\mu E + \frac{2\mu e^2}{r} - \frac{K^2 + H^2}{r^2}\right)}\,dr,$$

the energy levels turn out to be, not the usual fine-structure, but

$$E = \frac{-2\pi^2\mu e^4}{\hbar^2[p + \sqrt{(k^2 + l^2)}]^2},$$

with the appearance of three quantum numbers.
2. One thing that prevents this is 'phase invariance' – that is, that the complex function ψ describes the same physical situation as $\psi e^{i\alpha}$.
3. The following way of showing this is the most instructive. For a fixed volume V

of boundary S, the change in the content of V is produced by the flow across S:

$$\frac{\partial}{\partial t} \int_V \rho \, d\tau = - \int_S \rho \mathbf{v} \cdot d\mathbf{S} = - \int_V \operatorname{div}(\rho \mathbf{v}) d\tau.$$

This gives the identity, for any V,

$$\int_V \left[\frac{\partial \rho}{\partial t} + \operatorname{div}(\rho \mathbf{v}) \right] d\tau = 0,$$

which is equivalent to the usual differential equation of continuity

$$\frac{\partial \rho}{\partial t} + \operatorname{div}(\rho \mathbf{v}) = 0.$$

4. This way of describing the situation, which is the usual one, is somewhat generous to the Born interpretation. What is meant is simply that if, in

$$\frac{\partial}{\partial t} \int_V |\psi|^2 d\tau = \frac{\partial}{\partial t} \int_V \psi \psi^* d\tau$$

for a fixed volume V, one uses Schrödinger's equation

$$\frac{\hbar}{i} \frac{\partial \psi}{\partial t} + H\psi = 0, \quad -\frac{\hbar}{i} \frac{\partial \psi^*}{\partial t} + (H\psi)^* = 0$$

one derives for the derivative:

$$\int [\psi^* H\psi - \psi(H\psi)^*] d\tau = 0.$$

But H is always required to be 'self-adjoint', the definition of which is just that this integral shall vanish, or more generally

$$\int [\theta^* H\psi - \psi(H\theta)^*] d\tau = 0$$

for any ψ, θ. But if one asks for an equation of continuity of the form derived for a gas, one needs a definition of the velocity vector and this comes, as a mere tautology, as

$$\mathbf{v} = \frac{i\hbar}{2m} \frac{\psi \nabla \psi^* - \psi^* \nabla \psi}{\psi^* \psi}.$$

This arises as follows: From Schrödinger's equation and its complex conjugate one easily derives

$$\frac{\hbar}{i} \frac{\partial}{\partial t} (\psi^* \psi) + \psi^* H\psi - (H\psi)^* \psi = 0.$$

For a single free particle with $H = p^2/2m$, $H = -\hbar^2 \nabla^2 / 2m$ so that

$$\frac{\partial}{\partial t}(\psi^*\psi) + \frac{i\hbar}{2m}(\psi\nabla^2\psi^* - \psi^*\nabla^2\psi) = 0,$$

which can be rewritten as the equation of continuity

$$\frac{\partial}{\partial t}(\psi^*\psi) + \frac{i\hbar}{2m}\text{div}(\psi\nabla\psi^* - \psi^*\nabla\psi) = 0.$$

5. If the equation is written as

$$E^a\partial_a\psi + \frac{im}{\hbar}\psi = 0,$$

the vector $v^a = \psi^+ E^a\psi$, where ψ^+ is the complex conjugate of the transposed, is such that

$$\partial_a v^a = \psi^+ E^a\partial_a\psi + (\partial_a\psi^+\cdot E^a)\psi$$

$$= \psi^+ \frac{im}{\hbar}\psi + (E^a\partial_a\psi)^+\psi,$$

which is identically zero. Thus v^a is a conserved current. The same, however, is true of $w^a = \psi^+ E^{a5}\psi$. Independently of this, it is now generally (but not universally) doubted whether the Dirac equation has a consistent one-particle interpretation. But such worries were all in the future.

6. Essentially the notion referred to here is simply the well-known Newtonian reduction of the two-body problem to the one-body problem:

$$m_p\ddot{\mathbf{r}}_p = \frac{e^2}{r^3}(\mathbf{r}_e - \mathbf{r}_p) + m_p\mathbf{F},$$

$$m_e\ddot{\mathbf{r}}_e = -\frac{e^2}{r^3}(\mathbf{r}_e - \mathbf{r}_p) + m_e\mathbf{F},$$

where $\mathbf{F}$ represents external gravitational or inertial forces. Then adding gives

$$M\frac{d^2}{dt^2}\frac{m_p\mathbf{r}_p + m_e\mathbf{r}_e}{m_p + m_e} = M\mathbf{F}.$$

The quantity differentiated is the location of the centre of mass, $\mathbf{z}$ say, so that $M\ddot{\mathbf{z}} = M\mathbf{F}$. Next, dividing and subtracting,

$$\ddot{\mathbf{r}}_p - \ddot{\mathbf{r}}_e = \frac{1}{\mu}\frac{e^2}{r^3}(\mathbf{r}_e - \mathbf{r}_p)$$

so that, for the internal coordinate $\mathbf{r}$,

$$\mu\ddot{\mathbf{r}} = -\frac{e^2}{r^3}\mathbf{r}.$$

7. In fact it is just

$$H_e s + \frac{\hbar}{i} \frac{\partial s}{\partial t} = 0,$$

where H_e is the external Hamiltonian, the expression for the total energy in terms of momentum and so has the three terms $(p_x^2 + p_y^2 + p_z^2)/2m$.

8. This is simply $H_i s = E s$, where H_i has four terms.

9. For the radial direction the momentum conjugate to a displacement r is $(\hbar/i)(\partial/\partial r)$, whilst that conjugate to a displacement θ is $(\hbar/ir)(\partial/\partial\theta)$.

10. He actually employs a mathematical dodge to algebraicise this. He suggests the use, instead of r, of $\phi = \log r$. The corresponding conjugate momentum easily proves to be $(\hbar/ir)(\partial/\partial\phi)$; and interchange corresponds to ϕ increasing by $i\pi$. A continuous interchange will then be described by increasing ϕ by iq and so, just as the momentum conjugate to θ proved to be $(\hbar/ir)(\partial/\partial\theta)$, so here the momentum conjugate to interchange will be $-(\hbar/r)(\partial/\partial q)$, which is the extra term in the internal Hamiltonian. This argument is actually a carefully spelt out version of what has been clear from the introduction of interchange, earlier in the chapter, as essentially an angular variable. For angular variables like ϕ will always give rise to momenta of the form $(\hbar/i)(\partial/\partial\phi)$. The more detailed argument shows in addition that the factor r is the same for both ϕ and q, since they are variables of just the same kind.

11. It will be clear from the following simpler situation of a two-dimensional system, say a single particle moving in a plane. If the motion is described in terms of cartesian coordinates x,y, with conjugate momenta $p_x = (\hbar/i)(\partial/\partial x)$, $p_y = (\hbar/i)(\partial/\partial y)$, then the Hamiltonian equation $p_x^2 + p_y^2 = 2mE$ becomes

$$\frac{\hbar^2}{2m}\left(\frac{\partial^2 s}{\partial x^2} + \frac{\partial^2 s}{\partial y^2}\right) + Es = 0.$$

If now polar coordinates are used, with conjugate momenta $p_r = (\hbar/i)(\partial/\partial r)$, $p_\theta = (\hbar/ir)(\partial/\partial\theta)$, the classical Hamiltonian equation is again $p_r^2 + p_\theta^2 = 2mE$. However, mere substitution will produce

$$\frac{\hbar^2}{2m}\left(\frac{\partial^2 s}{\partial r^2} + \frac{1}{r^2}\frac{\partial^2 s}{\partial\theta^2}\right) + Es = 0$$

and this is *not* the result of transforming the cartesian Schrödinger equation to polars:

$$\frac{\hbar^2}{2m}\left(\frac{\partial^2 s}{\partial r^2} + \frac{1}{r}\frac{\partial s}{\partial r} + \frac{1}{r^2}\frac{\partial^2 s}{\partial\theta^2}\right) + Es = 0.$$

The latter equation must be the correct one. The failure to commute in this case shows up with the term p_r^2 which can be rewritten in many ways, one of which is $(1/r)p_r(rp_r)$. Classically the factor r can be cancelled but in the quantum form this is not so, and this form of the term produces the correct equation.

12. In fact, since F_4, F_5, F_{45} anti commute,

$$(C - H)^2 = (KF_5 + LF_4 + MF_{45})^2 = -K^2 + L^2 + M^2,$$

so that $C^2 = 2HC - (H^2 + K^2 - L^2 - M^2)$. Hence $C^2 = C$ only if either $H = 1$, $K = L = M = 0$, a trivial case, or $H = \frac{1}{2}$ and $-K^2 + L^2 + M^2 = \frac{1}{4}$, with the general solution $2L = \cos\theta\cosh u$, $2M = \sin\theta\cosh u$, $2K = \sinh u$. The simplest example is to take $C = \frac{1}{2}(1 \pm F_5 \pm F_4 \pm F_{45})$.

Eddington attached importance to E-numbers P such that, using the notation above $P^2 = P$ and $A = \frac{1}{4}$. Such an E-number is *factorisable* (that is, is the outer product of two vectors). The slight mystery with which he surrounds these results is dispelled if it is recalled that the E-numbers are the total 4×4 matrix algebra over C. The canonical forms under automorphisms for idempotent 4×4 matrices are well known to be

$$\begin{bmatrix} 1 \\ & 0 \\ & & 0 \\ & & & 0 \end{bmatrix}, \begin{bmatrix} 1 \\ & 1 \\ & & 0 \\ & & & 0 \end{bmatrix}, \begin{bmatrix} 1 \\ & 1 \\ & & 1 \\ & & & 0 \end{bmatrix}, \begin{bmatrix} 1 \\ & 1 \\ & & 1 \\ & & & 1 \end{bmatrix},$$

for which $A = \frac{1}{4}, \frac{1}{2}, \frac{3}{4}, 1$ respectively. The case of $A = \frac{1}{4}$ is self-evidently the factorisable one.

13. The general formula is straightforward to derive. Consider an infinitesimal parallelogram $ABCD$. Adopt geodesic coordinates at A so that $g_{ab} = \eta_{ab}$ and the first derivatives $g_{ab,c} = 0$ there. Take $AB = (dx^b)$, $AD = (\delta x^b)$ so that, to the first order, $BC = (\delta x^b)$ and $DC = (dx^b)$ as well (see Fig. 8.4). For parallel displacement of a vector A^b through dx^c, $dA^b = -\Gamma^b_{ac}A^a dx^c$. For displacement of A^b from A to B, D respectively, then, $dA^b = 0$, so that the displaced vector at B and D is also A^b. Now at B the coefficients of affine connexion have the values

$$\Gamma^a_{bc}(B) = \Gamma^a_{bc}(A) + dx^d\Gamma^a_{bc,d}(A) = dx^d\Gamma^a_{bc,d}.$$

Similarly at D,

$$\Gamma^a_{bc}(D) = \delta x^d\Gamma^a_{bc,d}.$$

For parallel displacement from A to C via B, then,

$$d_1 A^b = -\Gamma^b_{ac,d}dx^d\delta x^c A^a.$$

Similarly via D:

$$d_2 A^b = -\Gamma^b_{ac,d}\delta x^d dx^c A^a,$$

so that the difference

$$d_1 A^b - d_2 A^b = (\Gamma^b_{ad,c} - \Gamma^b_{ac,d})\delta x^d dx^c A^a$$
$$= R^b_{adc}\delta x^d dx^c A^a$$

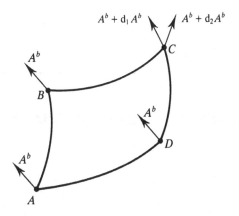

Fig. 8.4. Curvature defined by means of parallel displacement round a closed loop.

using the fact that $\Gamma^a_{bc} = 0$,

$$= \tfrac{1}{2}R^b_{adc}(\delta x^d dx^c - dx^d \delta x^c)A^a$$
$$= \tfrac{1}{2}R^b_{adc}A^a dS^{dc}$$

writing dS^{dc} for the element of surface area. For any curve bounding a surface S which can be divided into elementary parallelograms it follows that the change in a vector for displacement round the curve is

$$\delta A^b = \int_S \tfrac{1}{2}R^b_{adc}A^a dS^{dc}.$$

14. Such a recoil, since iq is added to ϕ, could equally be interpreted as imaginary change in ϕ or equivalently a complex gauge transformation in $r = e^{\phi}$.

If we are to pursue this method we must abandon Riemannian space and adopt Weyl's geometry which admits complex gauge transformations . . .. In wave mechanics we prefer to keep to Riemannian space.

This remark of Eddington's provides, then, additional support for a non-recoiling comparison fluid for interchange and also an argument (though an unnecessary one) that the physical consequences of inserting this non-recoil will be to give an alternative description of the same physical conditions otherwise described in Weyl's theory by a complex gauge transformation – that is, the electromagnetic field.

9

The proton–electron mass-ratio

The second of the two arguments of Eddington which I have chosen to analyse in detail occupies this chapter. It derives the quadratic equation whose roots are (approximately) the masses of the proton and the electron. In so doing it also fills the gap in the preceding chapter where the masses M, μ of the external and internal particles were assumed to have the values

$$M = \frac{136}{10} m_0, \; \mu = \frac{1}{136} m_0.$$

In other words, the result of this second argument is what allows Eddington in the preceding argument to dispense with any explicit counting of algebraic elements. It is not of much value to speculate on the historical details of the development of the two arguments. They were intertwined. The publication dates are all we have to go on. An inadequate argument on the charge on an electron (with 136 for 137) was published already in 1928 and a second attempt, essentially the argument described above, was included under 'Theory of Electric Charge' in 1932. A year before that a preliminary paper had appeared on the mass of the proton and a second attempt at this came out in 1933. This is substantially of the same form as in the book, which I describe in this chapter.

As I explained at the beginning of Chapter 8, the state of this second argument is much less satisfactory than of the first. I have therefore adopted a radically different approach. Whenever I feel that it is possible to replace Eddington's argument with a more satisfactory one, I have done so. In such cases I content myself with a short reference to the original argument. By the end of the chapter I shall hope to reach the situation of having an argument, more satisfactory than Eddington's but loosely based on the same general ideas. Because of this lack of historical validity, and also because the argument is a rather technical one, some readers may prefer to omit this chapter and proceed directly to the next one. This should occasion no difficulty in following my general argument about Eddington's writing of the two books.

Sources for m_p/m_e

In the last chapter I was able to show how the argument for the fine-structure constant originated in Eddington's experimentation with the Dirac algebra and the direct product construction. The origin for the argument for the proton–electron mass ratio is much more obscure and so I need to say something about it. One can get an important clue from a piece of Eddington's popular writing (Eddington 1933a) which was a working up of some public lectures he had given. This gives an informal description of the derivation of the quadratic equation on the following lines. This derivation is very tentative and significantly different from the version in RTPE which I am going to analyse in detail later in the chapter. It may seem very suspect, but it is none the less important for us to see how Eddington's thought was starting to develop. Eddington, talking to astronomers, begins with the notion of the radius of curvature, R, of an idealised universe (so, a space of constant curvature). Then he turns to the idea of an atom as treated in quantum mechanics with apparent disregard for the rest of the universe. He believes that, because of the practical correctness of quantum mechanics, R must be hidden somewhere in the description. But R is 'of the order of 10^{27} cm' whereas the conventional figure for the radius of the hydrogen atom is 10^{-8} cm. The ratio of these is a pure number of the order of 10^{35}. Eddington is now drawn straight to the 'large number argument' again. He estimates the total number of protons in the universe (the main source of mass) by dividing the astronomers' value for the total mass by the mass of a proton. The result is $N \approx 10^{79}$. Again, the ratio of the electrical to the gravitational force between proton and electron is about 10^{39}:

I have long thought that this must be related to the number of electrons and protons in the universe . . . and I expect the same view has been entertained by others. Since N is about 10^{79}, the above ratio is of the order of $\sqrt{N}$.

His next step is to give some very provisional arguments for $R/\sqrt{N}$ ('I do not profess to have achieved the necessary physical insight to settle the question ...') and then he notes with satisfaction that $R/\sqrt{N} \approx 3 \times 10^{-13}$ cm. He identifies this with the number called the 'classical electron radius',

$$\frac{R}{\sqrt{N}} = \frac{e^2}{m_e c^2}.$$

Since cosmology already gives, by conventional arguments in general relativity,

$$\frac{R}{N} = \frac{2\sqrt{3}Gm_p}{\pi c^2}$$

he is able to find R,N separately. The agreement with the observed values in 1932 was good but later observations have changed the picture completely.

The quadratic equation

The argument so far has been very much guided by the size of numbers, especially of large ones. There is still the possibility that various factors like 4π might need to be inserted. In addition, Eddington makes a change to what he calls 'quantum mechanical units' for mass. In these, he claims, $m_e c^2$ is given, not by $e^2 R/\sqrt{N}$ but by $(e^2/\hbar c)R/\sqrt{N}$. Then the equation seems to become

$$137m_e = R/\sqrt{N},$$

but Eddington argues, not very convincingly, for replacing this by

$$136m_e = R/\sqrt{N}.$$

Then comes the passage that illuminates the thinking by which he reached the quadratic equation:

But to this there is a serious objection. The result shows an unfair discrimination in favour of the electron, the proton not being mentioned. The proton is presumably as fundamental as the electron. But what can we put in place of $\sqrt{N}/R$ which would give an equally fundamental equation for the mass m_p of a proton?

With an electron and proton calling out for equal treatment the only way to satisfy their claims impartially is to make the fundamental equation a quadratic, so that there is one root for each. We do not want to alter the part we have already got, after taking so much trouble to justify it bit by bit; so we assume that

$$136m - \sqrt{N/R} = 0 \qquad \text{(C)}$$

gives correctly the last two terms of the equation, but there is a term in m^2 to come on at the beginning.

It is well known that we can learn something about the roots of a quadratic equation, even if only the last two terms are given. The ratio of the last two coefficients is the sum of the roots divided by the product of the roots. Since the equation is to have roots m_e and m_p, we must have

$$\frac{m_e + m_p}{m_e m_p} = \frac{136R}{\sqrt{N}}$$

or

$$\frac{136 m_e m_p}{m_p + m_e} = \frac{\sqrt{N}}{R} \qquad \text{(D)}$$

This is another change in the identification equation; but this time it is a very small change numerically. Comparing . . . we see that a factor $m_p \div (m_p + m_e)$ has been inserted. We know that m_p is about 1847 times m_e, so that . . . the change is insignificant; but the proton no longer has any cause for complaint, for proton and electron receive perfectly impartial treatment in (D).

The next step is to complete the quadratic equation of which the last two terms are given in (C). Since we have finished with the problem of the identification of the adjusted standard (our final equation giving it in terms of known experimental quantities being (D)) we may as well adopt it as our unit of length. As already explained this choice of unit ought to reduce the equations to their simplest possible form. This means that $R/\sqrt{N}$ can now be taken as unity. The two terms in (C) are therefore $136m - 1 = 0$, and the completed quadratic is

$$?m^2 - 136m + 1 = 0.$$

What number must we put in place of the query? You may remember that there was a number $n = 10$ which we promised to bring [in]. Here is our chance. We take the equation to be

$$10m^2 - 136m + 1 = 0.$$

Naturally one must beware of translating an ingenious popular presentation back into the original thought-processes. None the less, when the presentation is by the original discoverer, it must have some significance. The strikingly new feature of the argument is the intimate linking of the smallest and the largest scale phenomena. The atomic constants are related to the expansion of the universe in a huge extension of what Einstein had called Mach's principle. Mach himself had drawn attention to the fact that the local inertial frames of Newtonian mechanics were such that, relative to them, the most distant matter in the universe was not rotating. For

Einstein, this meant that the most distant matter determined, presumably by some causal means, the local inertial frames. He considered it a defect of general relativity that it did not 'incorporate Mach's Principle'. Here the further extension of the effect is to atomic phenomena.

Turning now to Eddington's more formal derivation of the quadratic equation in RTPE, there are two central notions involved. Both of these are given familiar names by Eddington: *phase space* and *combined system*, but although his ideas have some affinity with the conventional ones of the same names they are really very different. He does not always make this difference clear and my main task in this chapter is to lay it bare.

PHASE SPACE

Concepts of phase

The term *phase* has more than one meaning in physics, but all of them carry overtones of being a representation, which is not directly observed. The term *phase space* is a geometrical version of this representation but with an additional presumption of *equi-occupation*. By this I mean that the different possible states of the system are exhibited as points of a geometrical space and that the *a priori* probability of being at one point of the space rather than another is the same for all points. This refers to the probability before any mechanical constraints or equations of motion are imposed. The actual behaviour of the system will not have this uniform probability. Thus the physical input to the description is this modification of the probability distribution.

One type of phase space occurs in classical mechanics, where it is usually taken to be the geometrical space in which the motion is represented in terms of both position and momentum. For a particle moving in three dimensions, for example, the phase space is six-dimensional. In wave mechanics the term refers to the space over which the state function ψ or s is defined, so that for a quantum particle in three dimensions the phase space is also three-dimensional. It was this fact that gave rise to the initial misinterpretation of the Born picture of the Schrödinger theory, as I described in Chapter 8. The Born interpretation states that, if x,y,z are cartesian coordinates in this phase space, the probability that the representative point of the system of state-function s should lie in the small 'box' $dxdydz$ at x,y,z (where, as usual, dx denotes a very small increase in x

and so on) is $|s|^2 dxdydz$.[1] If cartesian coordinates are not used, this expression becomes more complicated, but in Eddington's view it is an essential assumption of wave mechanics that a set of cartesian coordinates exists for the system, whether it is convenient to use it or not.

These two different concepts of phase are not unconnected. There is a path joining them through statistical mechanics. This is a subject in which the classical mechanics of a large number of particles is considered – too large an assembly for it to be practicable to follow the fortunes of individual particles. There, instead of the state-function, there is a particle-density n, so that the number of particles lying in a small 'box' of volume dV is ndV. If N is the total number of particles, n/N is then a probability and so corresponds to $|s|^2$ in the quantum case. But it is part of Eddington's argument that the existence of cartesian coordinates is not essential (or even usually possible) in this case, so that here one cannot write formulae like

$$dV = dxdydz$$

everywhere.[2] Eddington's argument is that, at least approximately, the extra nine degrees of freedom in the E-number transformation

$$r \to r' = qrq^{-1},$$

after the six degrees of freedom of the Lorentz group have been removed, are to be identified with *both* of these senses of phase space. By demanding consistency in this double usage, his numerical results follow. One might almost call it a kind of mathematical pun, the more so because of the way in which Eddington uses as well the word *phase* in the quantum mechanical sense where the complex state-function s is written $s = u(\cos \theta + i \sin \theta)$ with u real. Eddington generalises this by replacing the single phase θ by an E-number and then saying (RTPE, p. 97) 'the space is a phase space in the other sense of the term phase'.

In keeping with these ideas, he takes the coordinates in phase space as angular ones. This fits in with the degrees of freedom in q because Eddington sees q as a product of factors of the form $\cos \theta + E_a \sin \theta$. Such a factor then gives rise to a displacement to θ (if θ is small) in the E_a direction. The notation fits the state-function definition equally well, because the phase has all the properties of an angle. On the next page he goes on to say

The volume of a ten-dimensional element of phase space is defined to be

$$d\omega = d\theta_1 d\theta_2 \ldots d\theta_{10}. \tag{7.421}$$

There is nothing wrong with this definition, though it must be said that such definitions are not arbitrary if equi-occupation is assumed, since a change in the definition alters the *a priori* probability. But there is a confusion with Eddington's later usage because he sometimes uses angular and sometimes length-like coordinates. The relation between them can be a matter of definition; one can write $x = R\theta$, where R is some fixed length. Often Eddington is concerned with a space of constant curvature and then R is naturally taken as the radius of curvature. Otherwise, R is just some length arising naturally in the calculation. A more natural volume element is then

$$dV = R^{10}d\omega = (Rd\theta_1)(Rd\theta_2)\dots(Rd\theta_{10})$$
$$= dx_1 dx_2 \dots dx_{10}.$$

A modified realisation of phase space

Eddington's use of phase space has been criticised by A. J. Coleman (Coleman 1945) from a point of view different from mine. Coleman sees the notion as central to the 'whole mathematical development' (of RTPE) and he notes that the numbers 10, 136 arise as dimensions of phase spaces. I shall return to Coleman's criticism below. Before that I want to explain what I consider to be the erroneous definition of phase space given by Eddington and to improve on it. The whole matter really goes back, as I said above, to the fifteen degrees of freedom of q in the E-algebra transformation

$$r \to r' = qrq^{-1}.$$

As I said in Chapter 7 (and more especially in Note 5 there) such transformations leave sums and products unchanged, so Eddington began by regarding them as changing the state-description r to a new 'frame of reference'. But changes of reference frame are expected to be Lorentz transformations (and rotations) with only six degrees of freedom in all. Eddington's problem was to understand the other $15 - 6$ degrees of freedom. He tries to do this by contrasting the transformation of different 'types' of E-number,[3] one type transforming differently from the above equation. Then he finds that there are certain qs which are such that the two formally different transformations are in fact the same. These qs form a set with six degrees of freedom. These are seen by Eddington as the six degrees of freedom of the Lorentz group. The remaining qs produce 'transformation generating phase space'. Pursuing his 'change of state-description' idea, he thinks of this phase space as representing changes in the internal configuration of the system. He emphasises the ten-dimensional nature of this phase space.

There is something a little wrong here, of course. There are really only $15 - 6 = 9$ degrees of freedom. I want to begin by clarifying this minor difficulty, so as to be able to count in the same way as Eddington in the later part of this chapter. The true situation is simply that it is convenient and permissible to treat the space as ten-dimensional and to use the extra degree of freedom to relate the different phase spaces which arise in considering a combined system. All that is happening is this: the straightforward definition (not Eddington's) of a phase space produces a curved space and the curvature at each point is fixed by the calculation. For technical reasons, Eddington is able to derive the results he wants by considering only a small neighbourhood of phase space and over this neighbourhood the space may be approximated by a space of constant curvature (the analogy in two dimensions would be the surface of a sphere). Now the phase space is only a convenient construct for representing the physical happenings, a picture, as it were. So it is allowable to enlarge this picture or to reduce it by changing the measure of length embodied in the constant radius of curvature. This possibility provides for an extra degree of freedom. It is therefore legitimate to treat the number of dimensions of phase space, as Eddington does, as $16 - 6 = 10$. There is also no difficulty in using this freedom to make the whole sixteen components of q relevant to the general case.

There is a much more serious defect in Eddington's formulation. It lies in the importation of the different 'types' of E-number, a distinction for which no physical significance exists.[4] It is quite difficult to imagine why Eddington has chosen this tortuous way to make his definition of phase space concrete unless it is that he was unable to carry out the algebra involved in the analysis of the automorphic transformation. I noticed in Chapter 7 that the way in which the Lorentz group was contained amongst the fifteen degrees of freedom was slightly more complicated than Eddington had realised. But the algebra can be carried out;[5] so that we have amongst the fifteen degrees of freedom just six which are the Lorentz group and nine others which represent, not change of reference frame, but change of system. Taking account of the conventional extra degree of freedom corresponding to the radius of curvature gives the ten-dimensional phase space to substitute for Eddington's.

Coleman's criticisms

I now discuss Coleman's objections to Eddington's theory. They are important because they were the only helpful and reasonable objections that Eddington received. He did establish a dialogue with Coleman in a way which he found impossible with his other, mostly older and more famous, critics. I then go on to consider whether these objections also affect the revised version of phase space that I have argued must replace Eddington's. Coleman (Coleman 1945) rightly describes Eddington's notion of phase space as the domain of probability distribution of a particle, and so it is by means of this concept that the physical interpretation of the theory is possible. This was a theme much taken up by Eddington after 1936, not as a result of Coleman's thesis, which appeared only in 1943 not so long before Eddington's death, but because he identified, not altogether correctly, a lack of care in this part of the theory as the cause of the intense criticism he encountered. Coleman draws attention to Eddington's hostages to fortune in three quotations from RTPE: on p. 5: 'it is an essential characteristic of (phase) space that it is occupied or has a finite *a priori* probability of being occupied'; on p. 97: 'phase space should be closed'; two pages on: 'it is important for our theory that the whole volume of phase space be finite'.

Coleman defines a different phase space from Eddington but the difference is not, as with my phase space, that the whole notion is realised in a different way but merely that another space is defined, mathematically completely equivalent to Eddington's.[6] In terms of this mathematically equivalent phase space Coleman then shows that Eddington's conjecture is wrong; the phase space is of infinite volume.

I do not believe that Coleman's result is fatal to Eddington's theory. Certainly it calls for some modification of the fairly straightforward arguments that Eddington gives. But all that is needed is precisely the same kind of treatment that occurs in conventional quantum mechanics when, because of excessive idealisation, say to plane wave solutions, the wave-function cannot be normalised but yet its values in two different localities measure the *relative* probabilities of being in one rather than the other. (Or, equally, the treatment which allows the use of plane wave approximate solutions for electromagnetic radiation, although the *total* energy in such solutions would be infinite.) Coleman himself makes two alternative suggestions to overcome the trouble. One is to adopt a different definition of volume, arranged so as to make the total volume finite; the other is to drop the assumption that all points of phase space are equally likely to be occupied. Then one could arrange that those parts contributing 'too much' volume are very unlikely to be occupied. Shortly before his death, Eddington wrote to Coleman that he favoured the second method; but it seems a counsel of despair since the whole virtue of phase space lies in equioccupation.

Coleman's objections apply, I think, with equal force to the quite different definition of phase space that I have given above. I have not investigated the total volume but the nature of the formulae[5] suggest to me an infinite volume. I leave the matter open, since I believe that the revised version of Eddington's argument that I am giving, as with Eddington's original, does not depend critically on the finiteness of the total volume.

Curvature of phase space

I turn instead to the other important feature of phase-space geometry, its curvature. The curvature of Eddington's phase space is treated in a cavalier fashion by both Coleman and Eddington. Eddington believed that he had given an 'inelegant' argument to show that the total volume is finite. This cannot be the case with a flat space and so it is reasonable to take it for granted that the space is curved. But when Coleman has shown the total volume to be infinite, the possibility of a flat space enters again. It will become clear that such a flat space would be a disaster for Eddington's arguments.

Fortunately Eddington is concerned only with the local properties of the phase space in the immediate neighbourhood of the origin and it is not difficult to compute the Riemann tensor there for my revised definition of phase space.[7] This shows the curvature to be non-zero. The phase space then has a definite curvature, although this can be given other values by a decision to draw the phase space larger or smaller, as I explained above. Such a curvature will naturally have consequences on Eddington's proposal to identify phase space with the uncurved phase space of wave mechanics. In fact this will be possible only approximately and only then under certain conditions. It is from these consistency conditions that Eddington's results follow.

I now follow Eddington's lead in the next two steps, which are just like those he takes with his erroneous version of phase space. The first step is to replace the phase space in the neighbourhood of the origin by the space of constant curvature best approximating to it there. The second step is to

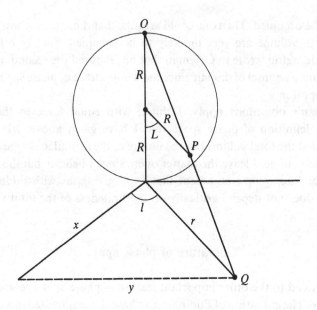

Fig. 9.1. Stereographic projection.

adopt for this approximate phase space a convenient coordinate system. Here the guidance is provided very much by the two-dimensional analogy, where a space of constant curvature is the surface of a sphere. The element of volume, which is important in using phase space, is here that of surface area. It is not convenient to use the geographer's coordinates, latitude and longitude for, as I mentioned in Chapter 3, small changes dL,dl in latitude L and longitude l produce a small displacement D of the sphere for which

$$D^2 = R^2[(\mathrm{d}L)^2 + \cos^2 L \cdot (\mathrm{d}l)^2].$$

As there are just the two squared terms and no product term, the effects of the two changes, dL,dl, are at right-angles. If they are taken separately, so as to form the sides $R\mathrm{d}L$,$R\cos L \cdot \mathrm{d}l$ of a small curvilinear rectangle, the area of this rectangle comes by multiplying the product of coordinate changes, dLdl, by the factor (called the *Jacobian*) $J = R^2 \cos L$. The factor $\cos L$ is the source of inconvenience, particularly as it vanishes at both the poles. A much more convenient coordinate-system is provided by stereographic projection (Fig. 9.1).

In this projection the sphere is imagined as standing on a plane with the South pole, S, in contact and every point P on the sphere (except the North pole) is projected onto the plane at a point Q by a straight line through the North pole, O. The ordinary cartesian coordinates (x,y) of Q in the plane are then used as numbers (*stereographic coordinates*) to specify the position of P on the sphere. It is clear that $SQ = r$ is given by

$$r = 2R\tan(45° - L/2)$$

and it is straightforward to verify that the small arc-length on the surface is now given by[8]

$$ds^2 = \frac{dx^2 + dy^2}{(1 + r^2/4R^2)^2}.$$

For stereographic coordinates, then, the Jacobian J is $(1 + r^2/4R^2)^{-2}$. The effect of the curvature is now just a conformal factor depending only on the distance r from S.

So much for the two-dimensional analogy. The corresponding theory in any higher number of dimensions was well known to Eddington. It gives a very similar result.[9] The metrical form is now

$$ds^2 = \frac{1}{(1 + r^2/4R^2)^2}(dx_1^2 + dx_2^2 + \ldots + dx_n^2),$$

where n is the number of dimensions and $r^2 = x_1^2 + x_2^2 + \ldots + x_n^2$ and so the corresponding Jacobian is $(1 + r^2/4R^2)^n$. Eddington proposes to regain the former angle coordinates θ_r by simply defining $\theta_r = x_r/R$, so that the metric form is simply

$$ds^2 = \frac{R^2}{(1 + \frac{1}{4}s^2)^2}(d\theta_1^2 + d\theta_2^2 + \ldots + d\theta_n^2)$$

where $s^2 = \theta_1^2 + \theta_2^2 + \ldots + \theta_n^2$ and the volume is

$$dV = \frac{R^n}{(1 + \frac{1}{4}s^2)^2}d\theta_1 d\theta_2 \ldots d\theta_n.$$

COMBINED SYSTEMS

Analysis

The second central idea in the derivation of the mass-ratio equation is that of a combined system. It is with this notion that Eddington seeks to make physical sense of the direct product construction of Chapter 7 which has inspired him by providing the sequence of numbers 3,10,136, ... It must be emphasised at once that Eddington's idea is intended to be quite a different one from the usual notion of combined systems in physics. The usual notion can be exemplified by the simplest classical system – a single particle moving in three dimensions. Its position can be described by three coordinates. A combined system of two particles, like the hydrogen atom, is then specified by giving the position of each particle and so needs $3 + 3 = 6$ coordinates. When Eddington considers the combination of two particles S and S' into a single system S_0 (RTPE, pp. 214–16) he says, apparently in direct opposition to this conventional view:

The phase space of S_0 contains 136 coordinates $\theta_{\mu\nu}$. It may be suggested that since the configurations of S and S' are each completely specified by ten coordinates, one of which they have in common, we shall require a phase space of 19 dimensions at most to specify the combined system S_0. But that is to begin at the wrong end of the problem. Our observational data relate to complex systems. We do not construct S_0 to represent data which have been ascertained about S and S'; we construct S and S' to represent data which have been ascertained about S_0. We have taken a system represented by a double wave vector in any combination of its 136 phases, all combinations being relativistically equivalent, and determine the conditions under which it can be dissected into two systems each with ten phases, or into a ten-phase system and a neutral comparison fluid with a single phase.

Observe the way in which the algebra dictates Eddington's analysis.

I think that a complete confusion has crept into Eddington's use of 'coordinate' and 'dimension' at this point. The conflict with the classical picture is much less than he imagines, for he is disregarding the curious process that gave rise to Darwin's '2 scalars, 2 vectors and a six-vector'. The crucial notion is that of the number of degrees of freedom of a complex system in four dimensions. The E-number algebra arose originally from Dirac's state-function which had four complex (so eight real) degrees of freedom. However this is expressed in E-number form, it is physically clear that the resulting E-number can have at most four complex degrees of freedom. Its sixteen elements must have some relations between them. To go back to a purely algebraic way of putting it, since Dirac's equation has by now been relegated to the background, we are concerned with the algebra generated by four elements E_a. This 'four' is indeed the number of dimensions of the system. The purely algebraic operations then give rise to $2^4 = 16$ elements.

The algebra for two particles

I have now to follow Eddington in generalising the idea of phase space to a system of two particles. I have to do this for the definition of phase space which I have constructed and which I argue is a corrected version of Eddington's flawed formulation. Before I do this, I need to look further at the actual process of separating off the two-dimensional sub-group from the eight, in the single-particle case. This is so that I may see how to generalise the method to the two-particle case.

In the one-particle case, then, one looks at the sub-group of the group of automorphisms,

$$r \rightarrow r' = qrq^{-1}$$

which is generated by the six elements $\cos \theta + E_{ab}\sin \theta$ for $a,b = 1,2,3$ and $\cosh u + E_{a4}\sinh u$. This sub-group lies in the eight-parameter sub-group for which

$$q = A + AE_5 + \tfrac{1}{2}H^{ab}E_{ab}.$$

Then one factorises out the transformations produced by $A + BE_5$. This is a process familiar in group theory where it is expressed abstractly. In the quaternion formulation[5] there is no need for this abstract treatment; it amounts to ignoring a factor which is a complex number. Why is this done (other than because it serves to derive the sought-for Lorentz group)? The answer is different for A and B. A may be disregarded because it cancels in the automorphic transformation equation. As to B, that is, to the phase of the complex number, such complex numbers produce what has come to be called a 'duality rotation' connecting corresponding components of the vector parts of r. The restriction to the Lorentz group is therefore equivalent to the forbidding of such duality rotations and there is a good reason for this. Under improper Lorentz transformations, like space reversal which turns right-hand into left-hand gloves, the behaviour of these two vectors is different. In modern parlance, one is a vector, one is a pseudovector and only one changes sign. It is therefore natural to forbid a transformation of one into the other.

Now consider the description of a combined system of two particles. The classical method is to go into eight dimensions, four for each particle (the particles being now treated relativistically). Eddington tries to realise this in terms of the direct product construction, so that his combined system inhabits the 'double frame' $E_\mu F_\nu$ with 256 degrees of freedom. It is natural that he should do this if, as I have argued, his original derivation of numbers came from this construction. In my own work in the past I have followed his technique (e.g., Kilmister 1951) but I now believe this to be wrong-headed. It does not fit very comfortably with Eddington's remarks quoted above about the analysis of a complex system being the logically prior step. It is not a matter of which algebra is being employed, for the double frame gives exactly the same Clifford algebra as results from taking the eight anti-commuting E_a, one for each of the degrees of freedom, in the obvious generalisation of the classical approach. It is simply a matter of how to go about the interpretation.

My disquiet at Eddington's double frame approach can be better understood by looking at the simpler case – of no physical importance – in the spirit of Eddington's argument about analysis. For a single particle moving in a plane, in classical mechanics, the two components of momentum can be taken as two of the components of a quaternion:

$$p = p_1 e + p_2 f$$

and the position similarly as

$$x = x_1 e + x_2 f.$$

As soon as one forms the product of these two quaternions, the rest of the algebra enters:

$$xp = -(x_1 p_1 + x_2 p_2) + (x_1 p_2 - x_2 p_1)g,$$

where $ef = -fe = g$ is the third quaternion unit. Both terms have a known significance classically and the more important one, the second, is the angular momentum about the origin. Further algebraic operations will inevitably bring in all four of the quaternion units together, but more will never be needed because the algebra is closed.[10] The combined system of two particles is described by the direct product of two quaternion algebras or, equivalently, by the E-number algebra. There are now sixteen degrees of freedom, but six of these correspond to changing the representation by Lorentz transformations and the remaining ten can be analysed as in Chapter 7. So a new entity is thrown up by the algebra, the 'combined system'. There is no clear and intuitive relation between the combined system and the two simpler ones into which it can be analysed. The analogy in Eddington's mind seems to have been the way in which the general physical properties of a hydrogen atom are not intuitively related to its analysis into a proton and an electron. That this was one model uppermost in his mind will be clear later, when he again introduces the two-particle transformation. This is not to say that such links cannot be forged; they are the results of the mathematical analysis. But they do not lie on the surface. They have to be worked for.

Now my argument here is that the analysis of the properties of the E-number algebra is done afresh in terms of the four anti-commuting E_a. The expression of these as products of the constituent quaternions plays no part in the interpretation. In the same way for the two-particle system the eight dimensions are to be realised by eight anti-commuting E_a, not by a direct product. To do otherwise would go against the spirit of the passage on p. 216 of RTPE about the analysis of a combined system. But of course the eight E_a correspond to the eight dimensions that would be the conventional way of describing a combined system. We arrange the numbering so that $E_1, \ldots E_4$ describe the first particle and $E_5, \ldots E_8$ the second. Following the analogy of the single particle argument above, the automorphism

$$r \to r' = qrq^{-1}$$

has now 255 degrees of freedom. Amongst the transformations those with

$$r = a + bE_{cd}$$

when $c,d = 1,2,3,4$, generate Lorentz transformations of the first particle and when $c,d = 5,6,7,8$ of the second. These form part of the subgroup of members of the form:

$$q = A + BE_9 + \tfrac{1}{2}C^{ab}E_{ab} + \tfrac{1}{2}D^{ab}E_{ab9} + \tfrac{1}{4}H^{abcd}E_{abcd},$$

with $1 + 1 + 28 + 28 + 70 = 128$ degrees   of   freedom.   Here $E_9 = E_1E_2E_3E_4E_5E_6E_7E_8$. A little more care is needed to determine the irrelevant degrees of freedom, for there are now more than those represented by A,B.[11]

In the first place, $E_1E_2E_3E_4$ corresponds for the first particle to what was called E_5 or i, and $E_5E_6E_7E_8$ is a corresponding element for the second particle, which I will call p. Then $ip = pi = E_9$. The group generated by $\cos\theta + i\sin\theta, \cos\theta + p\sin\theta$ and $\cos\theta + ip\sin\theta$ is then to be factored out on the grounds of producing duality rotations in one or both particles, as in the one-particle case. But that is not all. There is a remaining factor which has no analogue in the one-particle case. Although Eddington's phase space is different from mine, he gives a clue to this factor on p. 215 of RTPE:

To obtain the wave function of the combined system, the time coordinates t, t' of the two particles must be equated. Consider, for example, the sun–earth system. That does not comprise a combination of the earth today with the sun a week ago; no reference to the 'orbit' of such a combination will be found in astronomical textbooks. The essence of the process of 'combining' is the substitution of a single time coordinate for the whole system instead of independent time coordinates for its separate parts.

There are two algebraic elements connected with the time for each particle. I take them to be E_4, E_8 respectively. Then transformations with

$$q = \cos\theta + E_{48}\sin\theta$$

have the effect of changing the two time-variables:

$$t_1 \rightarrow t_1' = t_1\cos 2\theta + t_2\sin 2\theta,$$
$$t_2 \rightarrow t_2' = -t_1\sin 2\theta + t_2\cos 2\theta.$$

Suppose that, according to Eddington's requirement, $t_1 = t_2 = t$. Then $t_1' - t_2' = 2t\sin 2\theta$ and so Eddington's requirement is satisfied again only if $\sin 2\theta = 0$. What is important here is that the t_1 and t_2 are simply turned one into the other. One would not prohibit a transformation for which, for example, $q = \cosh u + E_{14}\sinh u$ merely because it changes the Lorentz frame in which the first particle finds itself. The important criterion is that there should exist *a* coordinate-system in which $t_1 = t_2$.

The consequent eight-dimensional group must now be factored out and so this leaves 120 degrees of freedom. This is the number 120 which Eddington had found by his various 'counting of algebraic elements' arguments. The present argument lifts it from that kind of numerology and identifies it with the number of possibilities in coordinate transformations. The dimensions of phase space in the two-particle case are therefore $256 - 120 = 136$.

The environment

I have now explained and modified Eddington's notion of phase space both for a single-particle system, where the phase space has 10 dimensions and for the combined system where it has 136. There is a related problem which he does not discuss. He is concerned with a combined system primarily because of the need he sees to consider a particle in its environment. The particle has a ten-dimensional phase space but the environment, if it is a simple isotropic one, has fewer. Indeed, Eddington takes its phase space to be one-dimensional; thus he says (RTPE, p. 219):

For an elementary particle and a neutral comparison fluid, the dimensions of the phase spaces are

$$n = 10, \; n_0 = 136, \; n' = 1. \tag{12.46}$$

(The strain vector of a neutral particle, being algebraic, has only one phase variable.)

A more cogent argument is that the environment has no particular 'positive' physical properties, only the 'negative' ones of being neutral, in the sense of producing no electrical effects, and isotropic, in the sense of having no preferred directions, but with one exception. The gravitational field cannot be wished away and it is determined by the Riemann tensor, that is, by curvature. So, this argument goes, there is just this one degree of freedom left. This will hardly do, however, for the curvature is described by the twenty components of the Riemann tensor. In a small region, however, one could approximate to the curvature by replacing the actual phase space by a space of constant curvature and then the radius of curvature is the desired one degree of freedom.

Neither argument is very convincing but the algebraic one suggests the following improvement. Reverting to the discussion of a spatially isotropic comparison fluid in Chapter 8, I showed there (partly in Note 12 of that chapter) that the general form

$$C = \tfrac{1}{2} + KF_5 + LF_4 + MF_{45}$$

could satisfy the condition $C^2 = C$ needed to allow the conventional ignoring of the environment on most occasions. Now the form given in Chapter 8 involved a C with two degrees of freedom. There is no question here of reducing these by means of Lorentz transformation; for the rotational part of the Lorentz group cannot change a spatially isotropic form and the effect of the translational transformations is to spoil the isotropy. But it is possible to simplify matters by the duality rotation, the F_5 transformation. This can absorb one degree of freedom, leaving just one as Eddington claims.[12]

Part 2   1928–33

THE PHYSICAL ARGUMENT

Consistency of probability

The physical argument to which Eddington applies the two notions of phase space and combined system falls into four steps (of which the first is ignored by Eddington). The situation considered is that of a particular state S_0 of a combined system, together with the particular corresponding states S,S' of the two constituent systems in the immediate neighbourhood of the first state. Each state belongs to a phase space of a certain number of dimensions – say n_0,n,n' for the sake of generality. These phase spaces are each replaced in the immediate neighbourhood of S_0,S,S' by spaces of constant curvature.[13] These may be equipped with stereographic coordinates, so that there is a metrical form for each:

$$ds^2 = \frac{1}{(1 + r^2/4R^2)}(dx_1{}^2 + dx_2{}^2 + \ldots dx_n{}^2),$$

where

$$r_2 = x_1{}^2 + x_2{}^2 + \ldots x_n{}^2.$$

In addition to this simplification for the whole of the respective phase spaces in the immediate neighbourhood, Eddington's takes account of the make-up of these phase spaces. The ten-dimensional space, for example, is made up of the eight dimensions of which the curvature has been found together with two more Euclidean dimensions, one of which is exemplified in the algebra by $\cos\theta + E_5\sin\theta$. In the combined system the corresponding Euclidean dimensions are four in number (for i,p,ip and the usual extra one from the conventional nature of phase space). In this case the dimension corresponding to kq has to be lumped with the other curved space ones because it does not commute with these. If the length measures are to be converted into angular ones, as Eddington often does, the corresponding constant lengths for the conversions will be R_0,R,R', the corresponding radii of constant curvature.

The first step in the argument, ignored by Eddington, arises in this way: one wants to compare the different Jacobians at corresponding points. But how does one decide which points, near to S,S', *correspond* to a particular point near S_0 of the combined system? Such a decision would need to go into the detailed connexion between the two single systems and the combined one, that is, into the details of actual problems. This would prevent the deriving of any results of some generality. There is a way out, and it is the assumption of the constant curvature approximation which provides it. The metric form is evidently isotropic in all the coordinates and, as I mentioned several times, the different phase spaces can be drawn with different scales. The scale is represented by the radius R. One can choose the scales so that, in the neighbourhood of S_0,S,S' corresponding points are equidistant from the origin, and so have the same value of r.[14] This assumes nothing about the way in which the correspondence is set up because it places only one constraint and that, one that can be satisfied by convention.

The second step in Eddington's argument is that, at such corresponding points, the probability function derived from particle number in statistical mechanics, which gives

$$P = pJ\mathrm{d}x_1\mathrm{d}x_2\ldots\mathrm{d}x_n$$

for the probability P of finding the system in a small box $\mathrm{d}x_1\mathrm{d}x_2\ldots\mathrm{d}x_n$, where J is the Jacobian,

$$J = \frac{1}{(1 + r^2/4R^2)^n},$$

should agree with the expression in quantum mechanics,

$$P = |s|^2\mathrm{d}x_1\mathrm{d}x_2\ldots\mathrm{d}x_n.$$

This will be so if $pJ = |s|^2$. Applying this to a combined system, Eddington notes two apparently inconsistent requirements:

(i)   The state-function for a combined system in quantum mechanics is given by the product $s_0 = ss'$, so that $|s_0|^2 = |s|^2|s'|^2$.
(ii)   Elementary probability requires that $p_0 = pp'$.

For these to agree, $J_0 = JJ'$ at corresponding points, so that

$$(1 + r^2/4R_0^2)^{n_0} = (1 + r^2/4R^2)^n(1 + r^2/4R'^2)^{n'}.$$

Such a condition could not be satisfied in general but in the neighbourhood of $r = 0$ it can be satisfied up to terms of order r^2 so long as

$$\frac{n_0}{R_0^2} = \frac{n}{R^2} + \frac{n'}{R'^2}.$$

This is really the basic equation but it has to be taken in conjunction with another, whose derivation forms the third step in the argument.

In this third step Eddington reverts to a structure of discussion which I described in Chapter 8. That is, he relies on the enlightenment thrown by striking models, without worrying too much about their mutual consistency. The model posed here is that of the two-particle transformation used in Chapter 8. In Eddington's use of it, it is applied to the two time-variables of the constituent particles but I think this is a distraction. What is more important in his discussion is the way he applies it to the Euclidean sub-spaces. He argues in this way: the aim is to derive results from a general abstract discussion of the combined system, but, as in the fixing of r, a certain amount of detail has to be brought in. He sees the relationship between the Euclidean dimensions as given by the two-particle transformation, giving rise to the coordinates x_0, y_0 in S_0 where

$$x_0 = \tfrac{1}{2}(x + x'), \, y_0 = x - x'.$$

For these coordinates it is easy to see that the partial Jacobians are such that

$$dx_0 dy_0 = dx dx'.^{15}$$

In terms of angular coordinates this gives

$$R_0{}^2 d\theta_0 d\phi_0 = (R d\theta)(R' d\theta').$$

But, Eddington argues, with these Euclidean dimensions one must have

$$d\theta_0 d\phi_0 = d\theta d\theta',$$

because these two expressions 'represent the same range described in different ways' (RTPE, p. 217). From this it follows that $R_0{}^2 = RR'$.

Eddington is emphatic that this condition is part of the whole condition imposed on volumes and he tries to prevent misunderstanding (RTPE, p. 217):

For example, in ten-dimensional phase space the volume element in local orthogonal coordinates $d\omega = d\theta_1 d\theta_2 \ldots$, contains an amount of probability $d\omega/\Omega$, where Ω is the whole volume of phase space which we have shown to be finite. Writing $d\omega_c$ and Ω_c for the corresponding volumes without the algebraic dimension, we have

$$d\omega = d\theta_{16} d\omega_c, \, \Omega = 2\pi\Omega_c, \tag{12.34}$$

and the probability $d\omega/\Omega$ is the product of the independent probabilities $d\theta_{16}/2\pi$ and $d\omega_c/\Omega_c$ that θ_{16} is in the range $d\theta_{16}$ and that the other coordinates are in the range $d\omega_c$.

Since $R_0^2 = RR'$, R_0 may be eliminated in the earlier condition on the radii, giving

$$\frac{n_0}{RR'} = \frac{n}{R^2} + \frac{n'}{R'^2}$$

and this is a quadratic equation in the ratio R'/R.

The fourth step in Eddington's argument is to relate this to the physical particles involved by asking how these phase-space radii will be recognised. The one property of particles which was then most evidently measurable and characteristic of particles was that of mass, as it still is. The other properties (in Eddington's time charge and spin, but now more numerous) have all a quasi-logical status. For example, to take the case of charge, elementary particles all have charges $\pm e$ or 0. I call this quasi-logical because the determination of charge requires the experimental answering of two yes–no questions. (The advent of quarks complicates the situation, but does not invalidate the argument.) Mass is different. All electrons have exactly the same mass and so do all protons. But apart from that, the mass seems to share more of the nature of macroscopic quantities with a range of values. If, then, one asks how a length R might be connected with a mass m, the answer in quantum mechanics is obvious. The dimensions of $\hbar/mc$ are those of a length, so that the connexion must surely be $R = K\hbar/mc$, where K is some numerical constant. The quadratic equation therefore becomes one for the mass-ratio m/m'. The two roots are the two values of the particles of mass m which can combine with the comparison particle of mass m' to give a combined system fulfilling the requirements laid down. Putting in the values for n_0, n, n' of 136, 10, 1 respectively gives the quadratic

$$10m^2 - 136mm' + m'^2 = 0.$$

Eddington makes a tacit appeal to his principle of identification to say that these two masses should be those of the proton and electron. With the usual notation, it follows from the theory of equations that the two masses satisfy

$$m_p + m_e = \frac{136}{10}m', \quad m_p m_e = \frac{1}{10}m'^2.$$

Reverting to the two-particle transformation employed in Chapter 8:

$$M = m_p + m_e = \frac{136}{10}m', \quad \mu = \frac{m_p m_e}{m_p + m_e} = \frac{1}{136}m',$$

which are the results quoted there (except that m' is now written for the comparison particle instead of m_0). The roots m_p, m_e have the ratio 1847.6. This ratio is not the observed value of the proton–electron mass-ratio, which is 1836.2 but it is near enough (within 0.6%) to make one think that there is a case for investigating Eddington's argument as closely as I have done.

It is natural to ask for some physical observation of the mass m' which is, very nearly, 136 (in terms of the electron mass as unit). None seems to be available. From a modern point of view it is noteworthy that the mass of the charged π-meson is very close to double this value (an error of 0.44%). But this is ahistorical. It was only two years after Eddington's publication of the quadratic that Yukawa proposed a particle whose exchange might account for nuclear forces (Yukawa 1935). He estimated the mass as between 100 and 200. The μ-meson, discovered in cosmic radiation the following year (Anderson and Neddermeyer 1936) was thought to be Yukawa's particle but it was found not to react strongly as it should. The matter was resolved after the Second World War (Lattes, Muirhead, Occhialini and Powell 1947) when the π-meson was found and identified with Yukawa's particle. So Eddington was not tempted into any identification of his m' with meson masses.

This chapter and the previous one have presented a somewhat modified version of the position Eddington had reached by 1933. It is not important for my general thesis to consider any of his other arguments in the same detail. What is already clear is that by 1933 he had reached a critical stage in his investigations. Results seemed to have confirmed his belief that the algebra introduced in physics by Dirac was the tool for which he had been waiting.

Notes

1. It is not essential to use cartesian coordinates. If spherical polars are preferred, for example, this expression becomes $|s|^2 r^2 \sin\theta \, dr d\theta d\phi$ so that a variable factor $r^2 \sin\theta$ relates the volume element with the product of coordinate differentials. And if a general set of coordinates with metric tensor g_{ab} is adopted, the corresponding factor is $\sqrt{g}$, where g is the determinant of the g_{ab}.
2. Eddington sometimes speaks as if there were an intrinsic connexion between curvature and $g \neq 1$ but he must have known otherwise. For it is hardly likely that he would be unaware of a view of Einstein (fairly quickly jettisoned) that one could and should choose coordinates in general relativity with $g = -1$

everywhere. The minus sign here is simply the result of the indefinite metric. This is certainly possible over an extended region. The true situation in Eddington's case is more complex, as will be seen later.

3. In fact he relates the automorphism in the text to the tensor transformation of a mixed tensor R_b^a of rank two. Thus

$$R_b^a \to R_{b'}^{a'} = X_a^{a'} R_b^a X_{b'}^b,$$

which, translated into matrix notation, with q written for $X_a^{a'}$, gives the automorphism. He then contrasts this with the transformation of a covariant tensor T^{ab}:

$$T^{ab} \to T^{a'b'} = X_a^{a'} T^{ab} X_b^{b'}.$$

This is rendered in matrix form as

$$t \to t' = qtq^t,$$

where q^t is the transposed matrix. Of course those qs which are such that $q^{-1} = q^t$, i.e. $qq^t = 1$, will be such that r,t transform in the same way. Such a set of qs is easily seen to have six degrees of freedom.

4. There is no physical difference between a mixed and covariant tensor; the only difference lies in the representation. This is put, in relativity, in the form of 'using g to raise and lower suffixes'. That is, one commonly writes there

$$t_b^a = g_{bc} t^{ac},$$

where g_{ab} is the metric tensor. The notion of phase space is too important physically to depend on transformation by the metric tensor.

5. As noted in Note 5 of Chapter 7, any q may be written as $q = a + kb$ where a,b are complex quaternions (though $i = E_5$ and $k = E_4$ so that $ik = -ki$). So long as q is non-singular ($|a| \neq 0$) it can be factorised into

$$q = (1 + kba^{-1})a,$$

so that

$$qrq^{-1} = (1 + kba^{-1})(ara^{-1})(1 + kba^{-1})^{-1}.$$

Now ara^{-1} is just r transformed by a Lorentz transformation. For present purposes, to investigate the effect of the rest of the qs, this transformation can be ignored. We are left with the phase-space generation:

$$r \to r' = (1 + kc)r(1 + kc)^{-1},$$

writing c for ba^{-1}, which is any complex quaternion. It would be laborious to compute $(1 + kc)^{-1}$, but this can be avoided by noting that

$$(1 + kc)(1 - kc) = 1 - kckc = 1 - c^*c,$$

using the fact that, since $ki = -ik$, $ck = kc^*$. Now $1 - c^*c$ is just a complex quaternion and so

$$(1 + kc)^{-1} = (1 - kc)(1 - c^*c)^{-1}$$

$$= \frac{(1 - kc)(1 - \bar{c}c^+)}{1 + |c|^2 |c^*|^2 - 2S.c^*c}.$$

The denominator is a real number. To the second order in c, which is all that is needed, it at once follows that

$$(1 + kc)^{-1} = 1 - kc + c^*c.$$

Writing $r = x + ky$, $r' = X + kY$, one is, in generating the phase space of r, to regard x,y as fixed and X,Y as variables determined by the parameter c. This easily gives

$$X - x = c^*y - y^*c - (c^*x^* - xc^*)c,$$
$$Y - y = cx - x^*c - (cy^* - yc^*)c.$$

Notice that, since c is a complex quaternion, this gives a set of points (X,Y) in sixteen-dimensional space determined by four complex, that is, eight real parameters. The expression seems to give an eight-dimensional phase space. The explanation is simple. The complex quaternion a by itself represents a Lorentz transformation. Multiplying a by any real number leaves the transformation unchanged. This removes one degree of freedom in the usual way. But now suppose that a is multiplied by a complex number z. Writing, again, $r = x + ky$,

$$X + kY = za(x + ky)a^{-1}z^{-1}$$

so that

$$X = axa^{-1}, \quad Y = (z^*z^{-1})a^*ya^{-1}.$$

The y-element is now multiplied by a complex number of unit modulus, and so the transformation is different. Summing up, the full ten-dimensional phase space is generated by the eight-dimensional formulae above and a two-dimensional aspect arising from z. The modulus of z can be used to represent the conventional radius of curvature of phase space and the phase of z provides a 'phase transformation'. If r is written in the usual form,

$$r = A + BE_5 + C^a E_a + D^a E_{a5} + \tfrac{1}{2}H^{ab}E_{ab},$$

the effect of

$$q = \cos \theta/2 + E_5 \sin \theta/2$$

is to leave A,B,H^{ab} unchanged but to mix the other two:

$$C^{a'} = C^a \cos \theta + D^a \sin \theta, \quad D^{a'} = -C^a \sin \theta + D^a \cos \theta.$$

6. Modifying Eddington's and Coleman's notations in line with Note 3, a mixed tensor R^{ab} transforms in matrix form to r', where

$$r' = qrq^{-1}$$

and a covariant tensor t goes to

$$t' = qtq^t.$$

Eddington's definition of the phase space at T is that it is the set of all E-numbers of the form

$$P_E = qtq^t - qtq^{-1}$$

for different qs. Coleman, on the other hand, sets up a correspondence between P_E and a new E-number P_C by the rule

$$P_C = qt^{-1}q^{-1}P_E + 1$$

so that

$$P_E = qtq^{-1}(P_C - 1).$$

The correspondence is therefore one-to-one and evidently the set of P_Cs is just as good as that of P_Es. But in fact P_C has a simpler form:

$$P_C = qt^{-1}q^{-1}(qtq^t - qtq^{-1}) + 1 = qq^t.$$

Coleman employs the set of P_Cs as phase space.

7. The expressions for the eight-dimensional phase-space displacements near the origin given at the end of Note 5 have the general form

$$r = c_i u^i + \tfrac{1}{2} c_{ij} u^i u^j.$$

Here the u^i are the eight real components of the quaternion c and r, c_i, c_{ij} are eight-dimensional vectors. It is straightforward to calculate the first derivatives of r and so the g_{ij} and then their first derivatives. From these one finds the Christoffel symbols in the usual way. In order to find the symbols of the second kind, it is necessary to know the values of the g^{ij}. It is only necessary to know these up to terms in u^i and by successive approximation one easily gets

$$g^{ij} = g^{0ij} - g^{0ia}g^{0jk}(c_{qp} \cdot c_k + c_q \cdot c_{kp})u^p$$

where the dot denotes the usual scalar product and g^{0ij} is the inverse matrix when $u^i = 0$, that is, inverse to $c_i \cdot c_j$. A detailed calculation then gives for the components of the Riemann tensor at the origin ($u^i = 0$):

$$R^a{}_{bcd} = c^a{}_c \cdot (c_{bd} - c_p c^p \cdot c_{bd}) - c^a{}_d \cdot (c_{bc} - c_p c^p \cdot c_{bc}).$$

It is straightforward to verify that this is not identically zero. The result has been found for the eight-dimensional phase space but of course the two extra dimensions cannot affect matters.

8. Clearly

$$\begin{aligned} dx^2 + dy^2 &= dr^2 + r^2 dl^2 \\ &= 4R^2[\tfrac{1}{4}\sec^4(45° - L/2)dL^2 + \tan^2(45° - L/2)dl^2] \\ &= R^2\sec^4(45° - L/2)[dL^2 + 4\sin^2(45° - L/2)\cos^2(45° - L/2)dl^2] \\ &= (1 + r^2/4R^2)^2 D^2. \end{aligned}$$

9. The only complication in more dimensions is in the specification of a space of constant curvature or, indeed, of the geometric curvature of a Riemannian space of n dimensions with Riemann tensor $R^h{}_{ijk}(h,i,j,k = 1,2,\ldots n)$. At any point P let u^i, v^i be any two vectors and consider all the geodesics through P whose tangent vectors at P have the form $ru^i + sv^i$. Such an array of geodesics

generates a two-dimensional surface through P determined by the pair of orientations u^i, v^i. Because the curves are geodesics, this surface is 'as flat as it can be' there. Its curvature K is defined by the Gauss surface theory. A detailed calculation then shows that

$$K = \frac{R_{hijk} u^h v^i u^j v^k}{(g_{hj} g_{ik} - g_{hk} g_{ij}) u^h v^i u^j v^k}$$

and this is independent of the orientations u^i, v^i if and only if

$$R_{hijk} = b(g_{hj} g_{ik} - g_{hk} g_{ij}),$$

where b is a scalar function of position. If this is differentiated once and Bianchi's identity (Note 8 of Chapter 3) is used, it follows easily that b is constant, giving Schur's theorem: If K is independent of orientation, then it does not vary from point to point. These two connected properties define a space of constant curvature. In such a space it is well known that there exists a coordinate-system in terms of which the metric is

$$ds^2 = \frac{k_{ab} dx^a dx^b}{(1 + Kr^2/4)^2} \text{ with } r^2 = k_{ab} x^a x^b,$$

the 'Riemannian form' for such a space. Here k_{ab} are zero if $a \neq b$ and when $a = b$ they are either 1 or -1 (according to the 'signature of the space').

10. Some may be tempted to use the scalar component of the x-quaternion for time but this would be wrong. That it must be wrong shows from the fact that such an identification would mean that $dt^2 + dx^2 + dy^2 + dz^2$ was invariant; and attractive as this might seem to a dishonest relativist prepared to insert some judicious square roots of -1, it is inconsistent with the invariant time-variable of classical mechanics. But I am sure that Eddington would argue that the reason that it is wrong is more subtle. If there is only one particle, there can be no time variable. For time, in classical mechanics, arises only when one compares motions. To include time, then, needs another particle and so a direct product of quaternion algebras.

11. An algebraic investigation can be carried out by the quaternion notation used in Note 5 above and Note 5 of Chapter 7. The general E-number $q = a + kb$ generating automorphisms can be sub-divided by considering first the case $b = 0$, so that $q = a$. This generates a sub-group of eight degrees of freedom but two of these are forbidden *viz.* the sub-group generated by $(1, i)$, the complex scalar multiples of q. Removing this removes two degrees of freedom leaving the Lorentz group. The forbidden part can be looked on as forbidden because it depends only on 1 and i and does not involve the quaternion units e_i. The elements of the 128-component q in the text can be analysed in the same way. To simplify notation I use as a temporary shorthand ab for E_{ab}. Then the first set of Es are describable by quaternions by the formulae

$$23, 31, 12 = e_1, e_2, e_3,$$
$$14, 24, 34 = -ie_1, -ie_2, -ie_3,$$

where $i = 1234$. Then if $k = 4$,

$$1,2,3 = -ike_1, -ike_2, -ike_3$$

and so on. The second set of Es similarly give

$$67,75,56 = f_1, f_2, f_3,$$
$$58,68,78 = -pf_1, -pf_2, -pf_3,$$

where $p = 5678$. Then, if $q = 8$,

$$5,6,7 = -pqf_1, -pqf_2, -pqf_3$$

and so on. In all of this we have to take into account the fact that i,k,p,q all commute with both the e_i and the f_i; i,k anti-commute as do p,q. It is easy to see that i,p commute and so do i,q and so by symmetry p,k. But k,q anti-commute. The 128 elements listed in the text are now written:

  (i) the two single terms $1, E_9 = ip$;
 (ii) the twenty-eight E_{ab}: e_i, ie_i, f_i, pf_i, $ikpqe_i f_i$, $kpqf_i$, $ikqe_i$, $kpqf_i$, kq;
(iii) the twenty-eight E_{ab9}: ipe_i, pe_i, ipf_i, if_i, $kqe_i f_i$, $ikqf_i$, $pkqe_i$, $ikqf_i$, $ipkq$;
 (iv) the seventy E_{abcd} which the reader may list.

The numbers which arise unattached to e_i or f_i turn out to be

$$1, ip, kq, ikq, pkq, ipkq.$$

The group is completed by i,p which turn up from multiplication and were already noticed in the text. The actual structure of this group is easily seen to be that of complex quaternions which, in this context, is better described as the Clifford algebra generated by three anti-commuting elements.

12. Note 12 of Chapter 8 derives

$$C = \tfrac{1}{2}[1 + F_5\sinh u + (F_4\cos\theta + F_{45}\sin\theta)\cosh u]$$

and the transformation

$$C \to C' = (\cos\phi + F_5\sin\phi)C(\cos\phi - F_5\sin\phi)$$

easily gives

$$C' = \tfrac{1}{2}\{1 + F_5\sinh u + [F_4\cos(2\phi - \theta) + F_{45}\sin(2\phi - \theta)]\cosh u\}.$$

This shows the dependence on u only, and choosing $\phi = \theta/2$, for example, gives a form making this explicit.

13. One procedure for choosing the closest fit space of constant curvature is as follows. At the given point, O say, one can choose a coordinate system ('geodesic coordinates') such that the metric tensor has diagonal form, with the non-zero elements ± 1 at the point and its first derivatives all zero. One can then calculate the Riemann tensor R_{abcd} there. Now we suppose that these coordinates are those of the standard form of a space of constant curvature at the point, with radius of curvature R (because the standard stereographic coordinates are indeed geodesic at the origin). This new metric has also a

Riemann tensor, depending on R, say R^O_{abcd} and this is a function of R. Then calculate $(R^{abcd} - R^{Oabcd})(R_{abcd} - R^O_{abcd})$ as a function of R and choose R so that it is a minimum.

14. Strictly r is not measured distance but it is so to the second order in r. For (i) along a coordinate-curve

$$ds^2 = \frac{dx^2}{(1 + x^2/4R^2)^2},$$

which gives $s = 2R\tan^{-1}(r/2R)$, or $s = r$ to the second order and (ii) a coordinate-curve is easily seen to be a geodesic up to terms of order r^2 since the coefficients of affine connexion for the metric have the form of linear forms in the coordinates, and so to the second order the geodesic equation is $d^2x^a/ds^2 = 0$.

15. This is because

$$\left| \frac{\partial(x_0, y_0)}{\partial(x, x')} \right| = \left\| \begin{matrix} \frac{1}{2} & 1 \\ \frac{1}{2} & -1 \end{matrix} \right\| = 1.$$

PART 3

1933–44

10

The turning point

The year 1933 marked the real turning point in the development of Eddington's thoughts. In this chapter I shall try to reconstruct his state of mind in that year, so as to assess the way in which he decided to go ahead with the writing of RTPE. My discussion up to now makes it possible to be more precise about the questions raised in Chapter 1. They can be broken down into six separate ones:

1. What made Eddington write RTPE?
2. Why is it so obscure?
3. What important and valuable aspects does it have?

Then 4, 5, 6 are the same questions for FT. It already seems likely from my argument, and it will prove to be the case, that the answers for FT are independent of those for RTPE. As a first stage the answer to 1 and in part to 2 will be found in this chapter.

The outside world

The last three chapters have been inward looking. I have not tried to relate the excitement that the world of physics was going through to the more pedestrian doings of the outside world. I think that is in keeping with the mood in Britain in the late 1920s. But it will not do for 1933. By then the world was showing sinister changes. In Germany Hitler had come to power in January. By the middle of the year the consequences were becoming clear. In particular the German preparations for a second world war and for attacks on Jews in Germany were seen to be in train. The consequent dreadful feeling of *déja vu* can be appreciated only by those at the time who had lived through the earlier war. To what extent was the quiet of Cambridge disturbed by this? More perhaps and sooner than in the country as a whole, where the press, especially *The Times*, played down the dangers. The news reached Cambridge in the more concrete form of those German scientists, mostly but not entirely Jews, who felt the need to leave Germany. Einstein, who had naturally been in occasional touch with Eddington, happened to be in the USA and never returned to Germany. By April a new law for 'reconstructing the civil service' removed Jews and socialists from university posts.

I can only speculate on the effect of this on Eddington. The coming war was in fact six years away and it did not seem unavoidable but the danger of it could not but distress a committed Quaker. As Herbert Jehle says (Jehle 1946):

Eddington stood for peace when it was hard to take such a stand. He wrote in 1940: I have found in the events of the last twelve months no ground for any weakening of my pacifist convictions.

One may contrast Eddington's standpoint here with the 'lesser evil' view of Russell. Russell's convictions of the wrongness of the First World War were changed into belief that entering the second was a lesser evil than refusing to do so. It seems to me that the distress engendered by Eddington's unwavering view, contrasted by the march of events, led already by 1933 to a change of heart about his work. It produced a desire to get something into some sort of finished form before disaster struck and interest evaporated, even if that meant that the beautifully polished style of MTR could not be attempted. Besides these external influences, Eddington must have been conscious of his age. He was now just over fifty. He seems to have been very fit, with no indication of the cancer that was to carry him off quite suddenly in eleven years time. But fifty is well beyond the age when most mathematicians do their best work. How much more acute this limitation seemed if there were to be a period of perhaps many years in which the world's interests were to be turned elsewhere.

Finishing RTPE

The external pressures were, as it happened, matched by internal ones. In the last two chapters I have explained the two loosely connected arguments of Eddington which had been finished by 1932. These arguments had introduced various concepts as well as the Dirac algebra that had started Eddington off on his theory and which he regarded originally as an extension of the tensor calculus. These new concepts were those of interchange, comparison, recoil and the combined system as well as Eddington's idiosyncratic formulations of indistinguishability, exclusion and phase space. I have not concealed my own opinion about the two arguments. They are intended to make important statements about the relation between quantum mechanics and the rest of physics. A by-product is the determination of two physical constants, one $1/\alpha = 137$, the other $m_p/m_e = 1847$. I believe that the first argument comes very near to fulfilling these intentions and the second is a good attempt. Though neither is acceptable in its entirety, there is enough right about them to serve as a basis for constructing correct physical arguments.

Eddington's opinion of the two arguments was doubtless more favourable than mine. In any case, even if incomplete, they provided something worth preserving. The arguments had already appeared in the form of papers by 1933. What had been the response? I have already mentioned the irascible Pauli's response to 137 and this differed only in intemperate statement from the majority view. Doubtless Eddington's colleagues and fellow astronomers were critical in conversation. But Eddington was no stranger to scientific controversy; astrophysics is one of the fields that particularly thrive on it. It is not likely that adverse critical reactions like that would have influenced him greatly. What was much more telling was the way in which his work was simply by-passed. Few papers made any reference to it. Scientists are a little like theatre people: it does not matter what the critics say so long as they say something. It may be significant that Eddington did not direct the attention of his research students to his new work, though it gave rise to problems in abundance. They mostly worked on technical problems in general relativity. Only the many-faceted George Temple, coming as a post-doctoral fellow, made early contributions to the algebraic theory.[1] It was ironic, from Eddington's point of view, that *his* further work on algebra in relation to physics, after he had gone to King's College London, was cut off by the need he saw to help the future war effort at the Royal Aircraft Establishment at Farnborough.

If these two arguments were insufficient to arouse response, what else was needed to overcome apathy? With hindsight, since the apathy resulted from the obscurity and logical shakiness of the arguments, the answer seems now to be to give better statements of the arguments. Eddington's reaction was different; his thought had already begun to move on in three different, though loosely related, directions which I shall describe in turn. Unlimited time would no doubt have allowed for these to be integrated together, and with the earlier arguments, to give a beautifully crafted sequel to MTR. Such time was not available, so RTPE became a report on 'work in progress'. He anticipates this change of technique, a major innovation for the disciple, in style, of Horace Lamb, in concluding a public lecture of 1932 which was published in an augmented form (Eddington 1933a):

> Now I have told you 'everything right as it fell out'.
> How much of the story are we to believe?
> Science has its showrooms and its workshops. The public today, I think rightly, is not content to wander round the showrooms where the tested products are exhibited; the demand is to see what is going on in the workshops. You are welcome to enter; but do not judge what you see by the standards of the showroom.
> We have been going round a workshop in the basement of the building of science. The light is dim, and we stumble sometimes. About us is confusion and mess which there has not been time to sweep away. The workers and their machines are enveloped in murkiness. But I think that something is being shaped here – perhaps something rather big. I do not quite know what it will be when it is completed and polished for the showroom. But we can look at the present designs and the novel tools that are being used in its manufacture; we can contemplate too the little successes which make us hopeful.

MTR and *The Internal Constitution of the Stars* had indeed been showrooms. RTPE was to be a workshop. Once we realise this we can see a clear set of reasons for Eddington's publishing of RTPE. Outside physics, the world situation and Eddington's age gave a sense of urgency. Inside physics, the lack of response to his first two arguments suggested to him the need for more. He had the bare bones of a second two, but their status was tentative. So the book was to be a workshop, not a showroom, an interim report on work in progress and it was this that was part of the reason for its obscurity. What it had to show, he considered, were the very satisfactory derivations of four constants; the fine-structure constant and the ratio of the masses of proton and electron came from the already published papers. A relation between large-scale and small-scale constants and a precise value for the constant N, the number of particles in the universe, came from two further arguments. There was also a first step to a correcting technique (the

Bond factor) to deal with the relation of simplified, exact theoretical results to complex experimental determinations. Eddington distinguished at that time seven 'fundamental constants'. The word 'fundamental' is somewhat question-begging. None the less, there would have been general agreement with his list:

The electron mass, m_e; the proton mass, m_p; the electron charge, e; Planck's constant, h; the velocity of light, c; the constant of gravitation, G; and the cosmical constant, k.

There was no question of finding all seven of these constants. Three had to be given to serve as a means of fixing the units of mass, length and time. The remaining four, Eddington felt, had been found.

The way ahead

In Chapter 8 I was able to show how Eddington gradually came to see that the fine-structure constant was within reach. In the last chapter I had to have recourse to Eddington's popular writing to suggest how he reached the quadratic equation for the mass-ratio. From 1933 to 1936 there is much less external evidence of Eddington's thought. He continued to publish on astrophysics some half a dozen papers but, with one exception, the remaining work for RTPE appears for the first time in the book. Accordingly some imaginative reconstruction is needed to establish Eddington's view of the best way ahead to finish the book. I have said that I distinguish three important driving forces. These were one, as in Chapter 9, from cosmology, one from Eddington's general philosophical approach and one internal to the work. I deal with these in turn. It may be thought that these arguments are confused and suspect, especially in comparison with the ones analysed in Chapters 8 and 9. Indeed they are, but they have to be negotiated to see how Eddington was thinking. Moreover, with the two earlier arguments, they will be found to have concealed in them the essential contribution that Eddington was to make in RTPE.

The inspiration from cosmology can be traced back to the controversy with Chandrasekhar and others on the 'relativistic degeneracy formula' although that formula plays only a very small part in the final version. The discussion into which Eddington had been forced by the controversy led him to question the automatic application of Lorentz invariance. Once he had begun to think in that way he began to see the possibility of cosmology providing a clear example of a situation which had only been hinted at before. This new idea is presented in the usual Eddingtonian way as a natural development of the theory although it is in fact a novel departure. I need to make clear the nature of this change in Eddington's intentions.[2]

Until 1933 the Dirac algebra was regarded by Eddington simply as an extension of the tensor calculus. The implication was that the tensor calculus, which had been adequate for general relativity, was just being enlarged to cover quantum mechanics as well. Such a view is a programme for producing a unified theory. The result of this programme would be a larger theory, parts of which would agree with simple relativity. The earlier chapters of RTPE confirm this position. For example, on p. 82, the fifth member of a pentad, E_5, is identified with the direction of curvature for a simple model of the world as a space of constant curvature. Eddington was entranced then by what I have come to think of as the 'magic' of the Clifford algebras and he continued to be so. I mean by 'magic' the perplexing way in which the algebra seems to provide a unifying *raison d'être* for a diverse collection of results.[3] The problem of tensor identities touched on in RTPE but worked out more fully in FT §89 shows Eddington's continuing fascination. This problem arises when one asks for the effect of certain symmetries on the breakdown of an *EF*-number into constituent parts (Kilmister 1951).[4] Later workers have also been powerfully affected by the Clifford magic. Two examples are (Chinea 1989) and (Grieder 1984). The Chinea approach is directed towards providing a convenient algebraic formulation of general relativity, whilst Grieder's aims at quantum field theory. These are two examples of the skilful utilisation of the technical advantages of the algebra. But it is very easy to step from that position to an erroneous one in which one sees the algebra as providing proofs of physical results otherwise unavailable. Eddington certainly comes dangerously near to making this step.

The algebraic approach can be seen either as a more convenient formulation of general relativity or of the existing tensor calculus, or perhaps as a modest extension of it but none of the work other than Eddington's has tried to exhibit it as providing a unifying link – other than a common notation – between general relativity and quantum mechanics. The common notation for the two is a gift horse not to be looked in the mouth but also not to be overestimated. The problem is not with notations but with the wholly different concepts of the two theories.

Multiple theories

The realisation of this led Eddington to a new point of view, which became fully clear to him in about 1933 and which forms the first of the three new directions to be incorporated in the later part of RTPE. The earlier view of the algebras as extensions of the tensor calculus which build a unified theory is very difficult to maintain alongside the new approach, though Eddington did so maintain it. This new approach was to regard his theory, partially built though it was, as destined to be other than a unified one. General relativity and quantum mechanics were to be allowed to go their own ways. Every theory, as any Kantian like Eddington would take for granted, is based on a particular approach which determines the categories under which the theory characterises nature. The rival sets of categories each provide a valuable insight. It might be that, in just a few simple problems, both approaches are applicable. Then there would be two apparently different descriptions. The agreement between these descriptions, Eddington thought, would depend on certain constants having appropriate values.

This approach is not based on extending the tensor calculus but much more on the model of Maxwell and the determination of the speed of light by comparing electric and magnetic quantities. The next step was to seek an instance of a physical situation which could be tried by both methods. The situation chosen by Eddington was the Einstein universe, which in general relativity is a very simplified static, homogeneous and isotropic model for the whole universe. This simple cosmological model had been known in general relativity for nearly twenty years.[5] Everything was known about this solution. It describes a fluid filling the whole universe with constant density ρ and pressure P where, in suitable units,

$$\rho = 3/R^2 - k, \ P = k - 1/R^2.$$

Here R is a spatial radius of curvature and (as in the previous description of this solution in Chapter 4) k is the cosmical constant.

Eddington's claim is to find a corresponding quantum mechanical description of the same physical system but there is much obscurity as to what this means. The details of the new approach appear on p. 203 of RTPE, a portion that is listed by Eddington as dating from the writing of the book in 1934–5:

It is important to realise that the individual atoms and electrons, whose energy and momentum are given by their wave functions, do not disturb the curvature of the space-time in which the wave functions are represented. Reciprocally the macroscopic objects, whose energy and momentum are represented by components of curvature of space-time, have no wave functions.

The first sentence must surely be taken to mean that the quantum mechanical description is to be found by solving Schrödinger's equation in the space-time of the metric of the Einstein universe but this cannot be correct.[6] It certainly will not do for Eddington's argument since it gives zero energy.

At the beginning of Chapter 14 of RTPE, however, in a different but also very clearly expressed description of the new method, there is a different meaning implied:

. . . it is evident that the curvature of space-time introduced in relativity theory and the waves of 'ψ' introduced in wave mechanics are equivalent. Both devices are used for the same purpose, to represent the distribution of mass and momentum of physical systems. Both are *devices*; it is not suggested that either the curvature or the waves exist in a literal sense . . .
But the result will be expressed in terms of different natural constants.

I pass over the extreme conventionalism of rejecting curvature as a mere device, unexpected from the author of MTR, and evidently inspired by a desire to show the symmetry of the two approaches. Instead I emphasise that here the curvature and the wave-functions are *alternatives*. The earlier suggestion which seemed to imply that Schrödinger's equation should be written in the given curved space has been dropped. Perhaps a calculation on the lines of that in Note 6 convinced Eddington that it was wrong. Now he proposes to compare the known results on the Einstein universe with the numbers derived from a wave-mechanical description of 'a self-contained static distribution of material particles without radiation'. For

I cannot but think that the realisation that a hitherto unrecognised relation exists – that there will be at least one redundant constant when the theories are brought together – is scarcely less important than the ascertainment of its precise numerical form.

He says in the preface to the book that this realisation was the last piece of the jigsaw:

after it was solved, there was not much difficulty in supplying the remaining investigations needed to fit together all the material.

Although Eddington offers to solve the Einstein universe by two methods, he never does this. Instead he proceeds by the same process as I have described at some length in Chapter 8, of putting together insights drawn from a number of models. But here the argument is conducted in a more excited manner, so that the distinctions between the models are much more obscure and, to my mind, impossible to disentangle.

A modified Eddingtonian attempt

Can one carry out Eddington's plan in a more straightforward way? It is clear that the free-particle Schrödinger equation will not serve for, if there is to be no curvature, the assemblage of particles cannot be self-contained. It is necessary to introduce a potential field V. The equations will then determine what value V will have at each point.[7] A detailed calculation shows that no solutions with any hope of physical interpretation can arise. Any hypothesis about the variation of $|s|^2$ with the radius r leads to a complicated V field. This must have become clear to Eddington fairly early on and so he disregards Schrödinger's equation completely and simply looks at a wave-function s which will correspond to the variation of proper volume in the Einstein universe. As he puts it:

Instead of treating R as a measure of curvature, we treat $1 + r^2/4R^2$ as a gauge factor. That is to say, we treat the stereographic projection as the true configuration; the curvature is abolished and x,y,z are rectangular coordinates in a flat space.

And to put a stop straightaway to any further suggestion of the need for the field V, he proceeds without further justification to calculating values at or near the origin.[8] This has the advantage of giving definite numbers but the justification for it is presumably that the origin in the Einstein universe is any point, so that what is calculated there is true overall. This argument is unfortunately circular for it is not now the homogeneous Einstein universe of general relativity (for which the contention is obviously true) with which we are concerned but the quantum mechanical assemblage. At the very least, homogeneity remains to be proved; and possibly it is untrue.

The calculation made by Eddington shows that near the origin the momentum operator $\mathbf{p}$ for a single particle in quantum mechanics is such that

$$\mathbf{p}^2 s = (9\hbar^2/4R^2)s,$$

or, in quantum mechanical jargon, 'the eigenvalue of $\mathbf{p}^2$ at the origin is $9\hbar^2/4R^2$'. Now a change of model is introduced. Eddington refers back to a standard treatment of an assemblage of electrons in a box, which he has given in a preceding chapter. This is the connexion of this argument with the Chandrasekhar controversy, for Eddington applied the standard treatment of electrons in a box to criticise the 'relativistic degeneracy formula'.[9] Using a result from that investigation,

$$\mathbf{P} = N\mathbf{p}^2/3m,$$

he gets $3\hbar^2 N/4mR^2$ as the eigenvalue of P. (Actually at this point he is still working with the case $N = 1$, so that he quotes $3\hbar^2/4mR^2$.) Then follows his variation of a standard quantum mechanical calculation:

To obtain the expectation value P, we must multiply the eigenvalue (at the origin) by the probability that the entity . . . is in unit volume at the origin . . . . Since the distribution is uniform, the probability associated with any unit volume is $1/V$ . . .. Hence

$$P = 3\hbar^2/4R^2mV, \quad \rho = 9\hbar^2/4R^2mV.$$

These results are in units for which $c = 1$. In general units the expressions are for P and ρc^2. The quoted results are also for the one-particle system. If there are N particles, the probability is N/V so that

$$P = 3N^2\hbar^2/4R^2mV.$$

The result for ρ is got from that for P by using a relation $\rho = 3P$, derived by Eddington by means of a rather curious argument.[10] If this same relation is used on the formulae for P,ρ in the standard Einstein universe, it gives $k = 3/2R^2$, so that

$$P = 1/2R^2, \quad \rho = 3/2R^2$$

in the units employed in general relativity. These units are so chosen that the constant of gravitation G is unity. The correct formulae in any units are

$$\frac{8\pi G}{c^4}P = 1/2R^2, \quad \frac{8\pi G}{c^2}\rho = 3/2R^2.$$

This is just the sort of result hoped for. Comparing the two expressions for P and using $V = 2\pi^2 R^3$, one gets

$$6G\hbar^2 N^2 = m\pi R^3 c^4.$$

Eddington does not deduce this particular relation between the macro-scopic G, N and R and the microscopic $\hbar, m$ but it is the relation that comes first in the attempt I have made to be more straightforward.

A question arises over the value to be given to the mass m. In Eddington's argument it is taken as the electron mass m_e but here it seems to me to arise very much as part of the background, so that m_0 or roughly $136m_e$, would be more appropriate. Taking as numerical values in cgs units (since these are always used by Eddington) $G = 6.66 \times 10^{-8}$, $\hbar = 1.04 \times 10^{-27}$, $c = 3 \times 10^{10}$, $m_0 = 1.2 \times 10^{-25}$ and accepting for the moment Eddington's value

$$N = 2 \times 136 \times 2^{256} = 3.15 \times 10^{79}$$

gives $R = 1.2 \times 10^{27}$ cm. Eddington quotes a value of $R = 1.2 \times 10^{27}$ cm which in his theory corresponds to a limiting value (the maximum) of the Hubble constant H for the recession of distant matter in the expanding universe of $H = 432$ km s^{-1} Mpe^{-1}. This was thought at the time to be near the observed value but all one can say is that the measured value should be less than this, which indeed it is.

Eddington's argument was more obscure than mine and his results are formally different though near to the same numerical values. He was evidently aware of the obscurity and strove to meet it in a way that he often did – by giving an entirely independent argument with similar conclusions. This is a very irritating practice, since it leaves the reader with the obscurity unrelieved. In this new argument, which he was to develop much more after 1936 he discusses the 'uncertainty of the origin' of the reference frame on the assumption that the origin is chosen as the centre of mass of all the particles in the 'background'. These particles are 'unspecified', so are at random (RTPE, p. 275) 'in the hypersphere of radius R which constitutes space'. The wording here is odd; space is the surface of the hypersphere and so it would be natural to write 'on' for 'in'. It is straightforward to work out the standard deviation of the distance of a particle from the origin; it is $2\sqrt{3}R$,[11] and so, by a standard result in statistics, for the centre of mass of N particles it is $2\sqrt{3}R/\sqrt{N}$. Eddington, taking his 'in' as correct, gets about $1/7$ of this, $\frac{1}{2}R/\sqrt{N}$. Eddington takes this as equal to the Compton wavelength of a background particle so that

$$\frac{2\sqrt{3}}{\sqrt{N}} = \frac{h}{m_0 c},$$

and therefore, taking the same value of N as before, $R = 2.9 \times 10^{25}$ cm which is of the same order of magnitude. This seemed to Eddington to confirm the 'one problem by both methods' hypothesis as a working tool.

Exactness and correctness

I turn now to the second of the three motivations for the work that completes the book. It is easiest to explain this in terms of an analogy quoted above. It is that of Maxwell's unification of electricity and magnetism. This determined the speed of light in terms of electric and magnetic constants. Now it is a matter of fact that the determination of the electric and magnetic constants was possible to a much higher degree of accuracy than the direct determination of the speed of light. Experimentally, then, there was not usually exact agreement. But once the theory had been accepted, the reaction to any discrepancy was never to doubt the equality of the derived and direct values for the speed of light, but always to attribute the discrepancy to experimental errors in the determination.

Eddington's thinking was more subtle than this Maxwell analogy, but it certainly incorporated it. More needs to be said about the nature of error and about the distinction, which seems to have become lost in Eddington's thought, between an exact theory and a correct one. The distinction is clear if one thinks of a digital clock; its reading of the time is exact but need not be correct. I will illustrate this by the two examples I have been treating in detail. The initial determination by Eddington of the fine-structure constant reciprocal was 136. He realised that this value was not correct and the search for an extra degree of freedom led him to the much more satisfactory theory I have described in Chapter 8. This gave rise to the value 137. This value is exact but it is not quite correct. The modern observed value is 137.036 (to three places). This is very accurately determined and is not likely to change very much in the future.

Even during the remaining years of Eddington's life it became clear that 137 was not the correct value, and he quoted 137.009 in 1944. His approach to this particular embarrassment was simply to question whether the experiments were really measuring what they claimed. Experiment is so complex that this is a very promising response but it needs a much deeper analysis of experiment than Eddington was prepared or able to give. None the less he made some attempt to improve some of his other numbers such as m_p/m_e but left the case of 137 to be shouldered by experimental shortcomings. The motive here seems to me to be to authenticate his theory by showing that it was correct – completely correct. Why should he make such a claim for a scientific theory, when no other would make it? I think it arises from the exactness of the predictions and the confusion between exactness and correctness.

In Eddington's attempts to make his results more correct a regrettably large part is played by the 'Bond factor' 137/136 which exceeds unity by 0.735%. This was suggested in (Bond 1935). The initial interest in this factor for Eddington lies evidently in the way it would turn the incorrect version 136 of the fine structure constant reciprocal into the correct 137 (as he saw it then). Bond's suggestion was actually more complex. He considered that the 'observational' values of e/m_e were erroneous, from the point of view of wave mechanics, by a factor 136/137. The real trouble with Eddington's use of the Bond factor is that it may occur raised to any power. Eddington chases through the determination of numerical values of constants in observational measurements to determine the correct indices on the Bond factors but his arguments are often obscure and sometimes clearly wrong.

My second example is one in which the Bond factor comes into its own. It is the quadratic equation. The original equation is

$$10x^2 - 136x + 1 = 0$$

and has the ratio of its roots 1847.6. If the correct result is to come from a quadratic, it must be one with slightly different coefficients. The quadratic

$$ax^2 - 2bx + c = 0,$$

where a, b, c are approximately 10, 68, 1 respectively, has the ratio r of its roots approximately $4b^2/ac$.[12] The result of small variations in the coefficients is given by

$$\delta r/r = 2\delta b/b - \delta a/a - \delta c/c.$$

It is possible to put the blame for the incorrect answer on any of the coefficients but Eddington's arguments directed his attention to the constant term, so that $\delta r/r = -\delta c/c$. Now the observed value for the mass-ratio in 1936 was 1834.1 and the latest observed value is 1836.16. The corresponding percentage errors in the original quadratic's prediction are 0.731% and 0.62%. Since the Bond factor exceeds one by 0.735% the current result in Eddington's time comes from simply replacing the constant term by the Bond factor, while the latest observed value results from using instead the Bond factor raised to the power 5/6.[13] Eddington gives arguments leading to the first of these modified quadratics in the later part of RTPE. When he came to writing FT in 1944 he argued instead for the second one. The obscurity of the argument, coupled with the freedom to introduce more or less any index, led to a general scepticism about the method amongst Eddington's readers.

The value of N

The third constituent of the later chapters of RTPE is Eddington's calculation of N, the 'number of particles in the universe'. At first sight it seems an outrageous claim to be able to calculate such a number. It might well be thought to be a contingent matter what value N should have in the actual universe. Eddington was well aware of this and he met such objections by taking up a conventionalist point of view (RTPE, p. 316):

Our ability to predict the number of protons and electrons in the universe implies that the number is imposed by the procedure followed in analysing the interrelatedness of our experience into a manifestation of an assemblage of particles or wave systems. It is a commonplace that electrons are not intrinsically distinguishable from one another; it is therefore not surprising that the total number, allowed for in our scheme of dissection of phenomena, depends on the conventional distinctions introduced when, for example, we decide that a certain diffuse wave packet is composed of two electrons rather than one.

Such an argument hardly faces up to the full extent of the difficulty. Eddington first defines $\frac{1}{2}N$ by dividing the astronomers' value for the total mass of the universe by the mass of a hydrogen atom. The factor $\frac{1}{2}$ arises because such a quotient corresponds to both protons and electrons. Such a definition is remote from the analysis into indistinguishables.

The actual form that Eddington gives the argument arises by considering, not the one number N, but a sequence

$$N_r = n_r(n_r + 1)2^{n_r^2}, \, r = 1,2,3,\ldots$$

where

$$n_1 = 2, \, n_r = n_{r-1}^2.$$

The first few values of n_r are 2, 4, 16, 256, . . . and so one easily gets

$$N_1 = 2 \times 3 \times 2^4 = 96,$$
$$N_2 = 4 \times 5 \times 2^{16} = 2 \times 10 \times 2^{16} = 1\,310\,720,$$
$$N_3 = 16 \times 17 \times 2^{256} = 2 \times 136 \times 2^{256} = 10^{79}.$$

Thus N_3 is close to the value for Eddington's N. The reason for the sequence is the possibility of first understanding better the argument for an imaginary world in which $N = N_1$ and then generalising.

This in turn rests on an argument going back to MTR (p. 323):

We can *observe* a relation between two physical entities. To *measure* a relation we must compare it with another relation of the same kind. Thus a measure is a relation between two relations and involves four entities . . .. The basis of measurement is therefore a four-point element of world structure. It is on this principle that I have developed the generalised field theory in *Mathematical Theory of Relativity*. Ultimately the theory of atomicity springs from the same origin.

Then on the next page:

The wave function is quadruple, since it specifies a quadruple probability distribution of four entities.

The idea here is that N_1 would correspond to an imaginary universe described by single wave-functions, that is, by E-numbers. Relations between entities then demand double wave-functions (EF-numbers). Finally these relations have to be set in an environment (comparison fluid) and this needs quadruple wave-functions or, if one were to need to express this algebraically, $EFGH$-numbers.

In the case, then, of simple wave-functions the number N_1 is said to arise in this way: the notion of a quantum particle involves an idempotent element of the algebra, that is, a P for which $P^2 = P$. Eddington argues that there are essentially sixteen such idempotents if the axes are chosen in the right direction.[14] A factor 3 arises from the fact that, in seeking an idempotent, the 'spin direction' may be chosen as E_{23} or E_{31} or E_{12}. The remaining 2 comes from the possibility of spin being positive or negative. This argument seems to me to contain an error. The figure sixteen arises by including, as well as zero and the unit operator, the six 'compact' E-numbers mentioned in Note 12 of Chapter 8 and ruled out by Eddington from representing particles. One is left with 8. I have set out the argument differently from Eddington but a similar defect was pointed out in his form of argument by Lemaître. In that context it takes the form of Eddington allowing an ambiguity of sign in a term which he had, earlier in the book, shown to be positive. Eddington's argument against Lemaître is not convincing. In modern terms it comes essentially to this: to each particle there is an anti-particle, so the number must be doubled. It does not in itself matter very much whether $N_1 = 96$ or 48 but this has implications later in the series.

For the cases of N_2, N_3 the argument is very similar and the factors 10, 136 are the dimensions of phase spaces as discussed in Chapters 8 and 9, corresponding to the algebras. But if, as Lemaître argues and I believe, $N_1 = 2 \times 3 \times 2^3 = 48$, then $N_2 = 4 \times 5 \times 2^9 = 10\,240$ and $N_3 = 16 \times 17 \times 2^{81}$ which is much less than N (in fact, near to $N^{\frac{1}{3}}$). So a question mark hangs over this calculation and Eddington carried out several others later on to meet these difficulties. One of these, already published elsewhere, is reproduced by the editor, Whittaker, as an appendix to FT. By 1935, however, Eddington was still confident in the correctness of the original form of the argument.

Publishing RTPE

Analysed in this unsympathetic way, Eddington's arguments may seem to have little claim on our attention. Yet each of these three directions of development gives an important indication of possible future work. The cosmological argument serves to emphasise the continuing role of Mach's principle and the need to bring the background into quantum mechanics. The Bond factor episode has unfortunate aspects but the positive side of it is the emphasis on the need for a second approximation to the crude values of the constants. And finally the argument for N exhibits the way in which the whole universe, not just the local environment, is involved. The truth of these points is perhaps arguable, but there is no mistaking the importance of the contentions if they prove to be true. Moreover, the spotlight has been turned on the finding of four constants; more of that particular list could not be expected. The sense of completeness engendered by this meant that Eddington felt able to begin the last section of the book with the triumphal:

Unless the structure of the nucleus has a surprise in store for us, the conclusion seems plain – there is nothing in the whole system of laws of physics that cannot be deduced unambiguously from epistemological considerations. An intelligence, unacquainted with our universe, but acquainted with the system of thought by which the human mind interprets to itself the contents of its sensory experience, should be able to attain all the knowledge of physics that we have attained by experiment.

Eddington's earlier published papers had met with some disbelief and a hostility that was largely concealed by his eminent standing. What, then, had the readers of RTPE come expecting? Probably in most cases an MTR sequel. The earlier book had taken a strange, largely uncomprehended theory and set it out clearly and elegantly. Nothing of the sort was to be found here. The disappointment led for the most part to harsh adverse criticism, criticism that Eddington felt unsympathetic and ill-informed. He reacted by ignoring it and turning in upon himself. There was only a little positive criticism, coming too late to influence Eddington, and I shall deal with this in the next chapter.

Notes

1. In fact he showed (Temple 1930) most ingeniously that the spectrum of the hydrogen atom could be found from the Dirac equation by purely algebraic means. It was not necessary to adopt any particular matrix form for the E_a nor (in the last paper of the four) need the momenta be interpreted as differential operators. All that was needed was the two sets of commutation relations,

$$E_a E_b + E_b E_a = 2\eta_{ab}, \; x^a p_b - p_b x^a = i\hbar \delta_b^a.$$

For Eddington's programme such a triumph of algebra was highly significant.

2. The date and nature of this change bears on a document in the appendix of (Douglas 1956) in which Eddington had provided a statement of his scientific achievements. Miss Douglas puts the date as between 1931 and 1937. The reference in the document to 'Eddington has therefore extended the tensor calculus . . .' seems to me to date it in the early part of that period, probably 1932 or 1933.

3. To take one example of the Clifford magic out of dozens, consider the differentiation of a bivector in special relativity. Write $H = \frac{1}{2}H^{ab}E_{ab}$ and $D = \delta^c E_c$, where $\partial^c = \eta^{cd}(\partial/\partial x^d)$. One has at once

$$DH = \frac{1}{2}\partial^c H^{ab}(\eta_{ca}E_b - \eta_{cb}E_a - \varepsilon_{cabd}E^{d5}),$$

so that, if $J = J^a E_a$ is the current vector, $DH = J$ combines *both* of Maxwell's equations in free space:

$$H^{ab}{}_{,a} = J^b, \; H_{ab,c} + H_{bc,a} + H_{ca,b} = 0.$$

The second equation shows that H is derived from a four-vector potential, but writing $H = DA$ gives *both* $H_{ab} = \partial_a A_b - \partial_b A_a$ and the Lorentz condition $\partial_a A^a = 0$ in the same way.

4. For example, consider an EF-number of the special form

$$kE_5 F_5 + T^{ab}E_a F_b + U^{ab}E_{a5}F_{b5} + \frac{1}{4}R^{abcd}E_{ab}F_{cd}.$$

If this is rewritten as a 'matrix with four indices', that is, as a mapping of tensors of rank two into tensors of rank two, and if certain symmetries are imposed on this tensor, the results are:

(a) The tensors T^{ab}, U^{ab} are symmetric, $T^{ab} = T^{ba}$, $U^{ab} = U^{ba}$.

(b) R^{abcd} has all the symmetries of the Riemann tensor,

$$R_{abcd} = R_{cdab}, \; R_{abcd} + R_{acdb} + R_{adbc} = 0.$$

(c) The sum of T^{ab} and U^{ab} is related to the contraction of $R^{abcd}, R^{bc} = R^{abc}{}_{d}$ by equations of the form of Einstein's field equations,

$$-(T^{ab} + U^{ab}) = R^{ab} - \tfrac{1}{2}g^{ab}(R - 2k).$$

(d) The contracted forms T, U are equal.

Of these, only (d) is without obvious significance.

5. In a frequently adopted coordinate system the metric form is

$$ds^2 = dt^2 - \frac{dx^2 + dy^2 + dz^2}{(1 + r^2/4R^2)^2}$$

where

$$r^2 = x^2 + y^2 + z^2.$$

6. The first problems in putting details into this programme arise in deciding how to define Schrödinger's equation in a curved space and how to define the state function s there. As to Schrödinger's equation, Eddington, with no argument, takes it for granted that the following process is involved:

(a) Observe that when curvilinear coordinates are used in a flat space in which cartesian coordinates are available, then the cartesian coordinates have to be used to 'quantise the system' (replacing p_a by $(\hbar/i)(\partial/\partial x^a)$) and the process of transforming the resultant Schrödinger equation simply replaces ordinary derivatives by covariant ones.

(b) If, then, no cartesian coordinates are available, simply write down the Schrödinger equation with covariant derivatives.

This process is plausible enough but it should be noticed that it excludes from the outset any genuine gravitational–quantum interaction. In such an interaction the RC tensor R^a_{bcd} would be involved and the process has ignored this. There is also a small technical problem to be clarified. That is the transformation of the state function, s, since that affects the form of the covariant derivatives. One instance of such a transformation under Galilean transformations, involving a phase change, was discussed in Note 7 of Chapter 5. Here the use of curvilinear coordinates raises a further complication. In cartesian coordinates $|s|^2 \mathrm{d}x \mathrm{d}y \mathrm{d}z$ is a probability of finding a particle in a certain volume. When this is transformed to curvilinear coordinates a Jacobian factor is introduced. Eddington always regards this factor as attached to the coordinate differentials and so for him $|s|^2$ is an invariant. Some people would prefer to attach the Jacobian factor to $|s|^2$, so that $|s|^2$ is a scalar density and s satisfies a slightly different equation.

Following Eddington's procedure, then, the Schrödinger equation for a single particle is derived by quantising $p^2/2m = E$ and is

$$-\frac{\hbar^2}{2m}g^{ij}s_{;ij} = Es,$$

where g^{ij} refers to the metric in Note 5. Writing $e^{-u} = 1 + r^2/4R^2$ and assuming that s is static and depends on r only (ground state) easily gives:

$$\frac{\hbar^2}{2m}e^{-2u}\left[s'' + \left(\frac{2}{r} - \frac{r}{R^2}e^u\right)s'\right] = Es,$$

where primes denote ordinary differentiation with respect to r. Now s will be complex, say $s = ge^{if}$ and the homogeneous assumption for the Einstein universe means that $g = $ constant, so that it is sufficiently general to write $s = e^{if}$. Dividing into real and imaginary parts easily gives:

$$-f'^2 = 2mEe^{2u}/\hbar^2, \quad 2f' + 2ru'f' + rf'' = 0.$$

The only eigenvalue is easily seen to be $E = 0$.

7. From $p^2/2m + V = E$ for a single particle, quantisation gives

$$\frac{\hbar^2}{2m}\nabla^2 s + (E - V)s = 0,$$

in which we again make the substitution $s = ge^{if}$ and assume that g, f and V are all functions of r only. This easily gives

$$g'' + 2g'/r - gf'^2 + 2m(E - V)g/\hbar^2 = 0,$$
$$gf'' + 2g'f' + 2gf'/r = 0,$$

where primes denote differentiation with respect to r. It is straightforward to integrate this to give $f' = B/g^2r^2$ and then g satisfies

$$g'' + 2g'/r - B^2/g^3r^4 + 2m(E - V)g/\hbar^2 = 0.$$

I have left the value of g unfixed; two possible suggestions would be $g = 1$ (uniform distribution, but then one that is infinite) or $g = e^{3u/2}$ ('self-contained' distribution with a falling off with distance that mirrors the Einstein universe proper-volume expression). Neither of these gives a V field making any physical sense.

8. s is proportional to $e^{3u/2}$ so that $s' = 3u's/2$ and $s'' = (3u''/2 + 9u'^2/4)s$. But $\nabla^2 s = (3u''/2 + 9u'^2/4 + 3u'/r)s$, so that, using $e^{-u} = 1 + r^2/4R^2$, $u' = -re^u/2R^2 = 0$ at $O, u'' = -1/2R^2$ at O gives the eigenvalue of ∇^2 at the origin as $-9/4R^2$ as Eddington gives. For $s = e^{ku}$ one gets the eigenvalue as $-3k/2R^2$.

9. Eddington's deduction is based on his identifying the quantised form of the classical energy tensor T^{ab} for N particles as

$$T^{ab} = N\mathbf{p}^a\mathbf{p}^b/m,$$

and then taking the generalised pressure operator as

$$\mathbf{P} = \tfrac{1}{3}(T_1^1 + T_2^2 + T_3^3) = N\mathbf{p}^2/3m.$$

This expression for P is just that from the kinetic theory where the pressure on a plane $x = $ constant is $P = \langle Nmv_x^2\rangle$, and $\langle v_x^2\rangle = \langle v_y^2\rangle = \langle v_z^2\rangle = \tfrac{1}{3}\langle v^2\rangle$, giving

$P = \langle\frac{1}{3}Nmv^2\rangle$. However, Eddington's justification for the form of T^{ab} is that the corresponding unquantised form is

$$T^{ab} = Nm\frac{dx^a}{ds}\frac{dx^b}{ds}.$$

10. The contracted energy tensor

$$T = T_a^a = T_{44} - T_{11} - T_{22} - T_{33}$$

gives

$$\rho_0 = \rho - 3P,$$

where ρ_0 refers to rest mass. This is then reinterpreted as a division of energy into a part described by standing waves and a part 'taken care of by the external wave function':

But here the whole curvature of the Einstein universe has been replaced by wave functions . . . . Thus the density $\rho = 3P$ . . . is the whole of the density.

11. Evidently

$$\langle x^2\rangle = \frac{1}{V}\int\frac{x^2 dxdydz}{(1 + r^2/4R^2)^3}$$

and so

$$\langle r^2\rangle = \frac{1}{2\pi^2 R^3}\int\frac{r^2 dxdydz}{(1 + r^2/4R^2)^3} = \frac{2}{\pi R^3}\int_0^{\infty}\frac{r^4 dr}{(1 + r^2/4R^2)^3} = 12R^2.$$

12. The ratio is $r = (b + D)/(b - D)$, where $D^2 = b^2 - ac$, so that D is approximately $b(1 - ac/2b^2) = b - ac/2b$, and r is $2b/(ac/2b)$ as stated.

13. $0.62/0.735 = 0.8435$ which has continued fraction form

$$\frac{1}{1+}\frac{1}{5+}\frac{1}{2+}\frac{1}{1+}$$

which shows that 5/6 is quite a close approximation.

14. To shorten Eddington's roundabout argument one can use the canonical forms in Note 12 of Chapter 8. There are respectively 4, 6, 4 and 1 of the various forms listed, making 15 in all or 16 if the zero is included.

11

Critical views of RTPE

The adverse criticisms of RTPE centred on two aspects – its obscurity and its claim to calculate physical constants. I shall deal with these two aspects in turn.

Obscurity

One reason that RTPE was found obscure by its readers, I have argued, was that they expected something different in style. But there was a more important reason; they also expected something different in content. For, though they had read MTR with approval, they had not appreciated the novel philosophical positions Eddington had taken up already by 1923. These positions arose out of general relativity but the main lesson of Eddington's book was, as its name implies, the mathematics of the theory. This could be read and the philosophical standpoints ignored.

I distinguished eight novel ideas that Eddington exhibited in MTR. All are relevant to RTPE and I discuss their relevance here in three groups of increasing importance:

(i) The notions of selective subjectivism and of falsifiability are neither of much importance, though for different reasons. It is hardly surprising that selective subjectivism fits well, for it is consciously followed by Eddington and expresses his Kantian preoccupation. Falsifiability is not important because it has taken a back seat. Eddington allows himself such freedom to change to a new model half-way through an argument that the ostensible falsifiability of an exact numerical prediction is a hoax.

(ii) Intermediate in importance are non-redundancy, descriptive tolerance and structure. Non-redundancy is taken for granted and used without
mention from time to time. For example, the fact that just four E_a give rise
to a fifth E_5 which anti-commutes with all the four gives rise to various
attempts at interpretation. It is taken for granted that there must *be* an
interpretation and that it will be physically important.

Descriptive tolerance is used frequently, almost always in the form:
general relativity (or cosmology) does it this way, quantum mechanics does
it that way and each regards the other as mistaken. We must accept both
and compare.

The notion of physics as structure is implicit throughout in this form: the
whole theory is engendered by various related linear algebras. The
algebraic structure is the structure of the physics but, because algebra is
essentially a study of structure, this fact writes in the notion of physics as
structure from the beginning. This is an idea that would hardly cause a
ripple of dissent now; but in 1936 it was an unusual view. In MTR
Eddington had perceived a possible limitation on this structural view:

> The possibility of the existence of an electron in space is a remarkable phenomenon
> which we do not yet understand.

It is the algebraic structure that takes over the task of including that
phenomenon as well.

(iii) The utmost importance is to be attached to the ideas of operationalism, the principle of identification and theory-languages. Operationalism,
in Eddington's sense, owes a further debt to Kant. But another aspect
which was mentioned in MTR can be seen to have come into its own in
RTPE:

> Any operation of measurement involves a comparison between a measuring
> appliance and the thing measured. Both play an equal part in the comparison.

There are various uses of the principle of identification. For example, when
the quadratic for the mass-ratio is found, the principle is used to identify the
masses as those of the proton and electron. Again, in one of the
interpretations of the E_5 term, curvature is imported into the discussion
because the term behaves just as it should to represent curvature.

Most important of all, however, in understanding RTPE is the implicit
use of the idea of a theory-language. This is especially needed in the form:
that there will be different theory-languages of differing complexity and it
would be wrong to use a more complex language to criticise results in a

simpler one. As I explained in Chapter 4, Eddington never formulated this idea, though it is implicit in his arguments through more than half of RTPE. It is, however, at variance with the Bond factor method of correcting results.

It is now possible to judge the reason for the adverse criticism of RTPE. It was in part justified by mathematical errors and by the unexpected workshop character of the book. But only in part, there was also a failure to appreciate Eddington's philosophical position, although that position could have been seen from a careful reading of MTR. This answers the second of our six questions. The reception of RTPE as 'obscure' was accepted by Eddington. This can be seen from an unpublished manuscript of a talk he gave. The date is uncertain, but I would guess 1938 or 9. In it he describes solving the Einstein universe by two methods. He says

I shall not be able to give you in this address sufficient detail for you to check the accuracy of my solution. But I will put before you some of the principal considerations involved – sufficient, I hope, for you to see that it is a quite tractable problem so if, when you come to read my full solution you are dissatisfied, or even (as I have heard whispered) found my treatment obscure, it does not matter so very much. I am not specially anxious to convince you that I have solved this problem. What I want is to convince you that *you* can solve the problem – so that anyway physics will not long continue to be retarded for lack of this fundamental relation of the constants.

Osborne's criticism

I turn now to the third question, that of the lasting value of RTPE. I deal with this by first considering other critical studies and then I deal with my own. It is pointless to go into detail with what were essentially ill-informed criticisms of the book. But one piece of valuable criticism came in 1944 from M. F. M. Osborne, at that time working at the US Naval Research Laboratory in Washington. This came too late to influence Eddington's thinking but the criticism is useful in its own right. The fullest exposition of Osborne's views is in a mimeographed note, the relevant parts of which are as follows.

Osborne believes that RTPE can be understood but not in terms of conventional ideas. One has to understand Eddington and the spirit in which he wrote. This is pretty near to the point of view that I have taken; that the eight original Eddingtonian notions need to be accepted first.

Next Osborne concludes, from internal evidence, that RTPE was written by the 'notebook' method: that is, that Eddington kept a notebook and wrote ideas as they came to him on every other page. At the end, guesses Osborne, he formulated general principles and then wrote out on the blank pages the arguments supporting the statements on the opposite pages. I am sceptical about Osborne's conclusions here, although his hypothesis does fit the facts of the book. It is also in general agreement with Eddington's 'workshop' description of his own work. But a 'notebook' method is completely at variance with what we know about Eddington's manuscripts for FT, as will be explained in the next chapter. None the less, I quote Osborne's further examples of incidents tending to support the 'notebook' hypothesis:

> There are often arguments given for two contradictory statements, without any attempt at resolution, for example, those for 136 and 137 as the value of the reciprocal of the fine-structure constant.

> He often introduces new ideas without developing them or ignoring the fact that he will later show them to be erroneous. Thus on p. 40 an overbold statement about the $q(\ )q^{-1}$ transformations is modified (or should be, but the discrepancy is ignored) on p. 125 when differential operators are introduced. Similarly, the volume of phase space is sometimes hopefully finite, sometimes admittedly infinite.

> Again, results to be proved later are often introduced. On p. 6 this is stated clearly:
>
> So with most of the physical conceptions; we have to introduce preliminary notions before the theory is sufficiently advanced for a full treatment.

> On pp. 120, 144 there is the assumption and use of a constant m, when the manner of its introduction means that it could be variable. On p. 231:
>
> The foundation of this exclusion principle will be investigated in Chapter XVI. Meanwhile we accept the general idea of such a principle from current quantum theory.

> Although on p. 119 there was the warning
>
> It would be foreign to our plan to intermingle the current semi-empirical theory with the purely deductive theory that we are developing.

Osborne also alerts us to three further difficulties in understanding RTPE:

> Everything is written from Eddington's point of view as if it were the accepted and familiar one and this confusion is aggravated by his use

of familiar technical terms to mean different concepts from their usual meaning.

There is a certain amount of 'downright cheating'.

He is prone to calling on his (undoubtedly great) physical intuition, or on some new principle, in the course of a solution.

I have given examples of the first and last of these in my discussion. It is harder to be sure about the cheating; what is cheating to one man is a provisional use of physical intuition to tide over a gap in the argument to another.

Osborne next sets out what he sees as a set of basic principles which Eddington is using:

Observer participation and the principle of relativity (not in the usual sense, but as needing two objects and then measurement needing two such pairs). Osborne gives a good example of the need for a comparison particle in conventional theory – Bohr's proof that the magnetic moment of the electron cannot be measured, by means of separating two beams of electrons of opposite spin, except in the presence of the nucleus. (The argument is an application of the uncertainty principle.) But Eddington claims that a similar situation is true for everything and if no comparison is stated, the background (Einstein) universe is implied.

The notion of a *combined system*, especially in the form of requiring a product rather than a sum of attributes.

The principle of unconscious approximation: Eddington claims to have a correct theory. Current physics is an (unconscious) approximation. To get the most agreement, some constants (the fine-structure constant, the proton–electron mass-ratio, N) must have appropriate values. (Examples given of this by Osborne, not from Eddington, are the electrical engineer's value of 377 Ω for the impedance of free space, the fact that G has the same value in Newtonian gravitation as in general relativity, and the history of electron spin.)

The principle of observer disturbance (which is more general than the uncertainty principle) and *the uncertainty of the origin*.

After some sections on the actual mathematics of the book and on specific Eddingtonian arguments, Osborne assesses the physical applications:

In general Eddington is marvellously skilful and provocative in stating a physical problem. In fact, this is perhaps the most attractive part of the book. He can also state with a modicum of lucidity what he is going to do in order to solve it. However, when a solution takes place it is often so obscure and so shot through with intuitive interpretations of the mathematics, as to leave the reader with a feeling of futility and frustration.

The examples that Osborne gives are mostly those I have discussed in earlier chapters.

Scale constants

I turn from Osborne's comments, which are directed to reducing the obscurity, to the other feature of RTPE which caused great suspicion: the calculation of physical constants. I said in Chapter 1 that it is a little hard to understand why the reaction to this was so strong. The aim of theoretical physics is always to calculate numerical results of experiment; the criticism of Eddington was really no more than that he had started from too narrow a base – had been a little too successful, as it were. He was partly to blame over this, for he expressed himself unguardedly. In (Eddington 1939) (p. 25) he says:

In physics *a priori* conclusions have long been anathema . . .. It has come to be accepted as a scientific principle that we can have no *a priori* knowledge of the universe. Agreed: provided that by 'universe' is here meant 'objective universe', as was undoubtedly intended when the principle was framed. But, as applied to a universe defined as the theme of physical knowledge rather than by its intrinsic characteristics, the principle cancels itself.

Eddington seeks by this means to insulate himself against the *a priori* accusation. He does this at the expense of having two universes, an objective one and one defined as the theme of physics. Such a dualism naturally brings even greater problems in its train.

A better reply would have been to point out that the conclusions should not have been described as *a priori*. A considerable amount of physical content has gone into the assumptions. For example the Dirac algebra starts with four E_a not three or five; and it is related back, in Eddington's theory, to the tensor calculus in four dimensions. Interpretations of algebraic quantities as momentum and energy are tied to the latter quantity arising as the square of the former. And so on: a variety of physics is assumed at the beginning. It is surprising, but not obviously impossible, that from such a beginning other numerical results should follow. None the less, a strong scepticism that the results were 'only numerology' began to get around. By this is meant a construction of numbers which is based mostly or solely on knowing the answer.

Although the scepticism about Eddington's calculations was misplaced, there was a deeper difficulty which seems not to have been much noticed at the time and which I shall show later to be the key to understanding RTPE. As he came to write RTPE, Eddington seems to have begun to realise that the problem of calculating constants divided into two parts. It was practicable, to put it no more strongly, to formulate a structural theory that gave rise to a certain small set of numbers: 137, 1847, 10^{79} and a few more. These numbers are dimensionless and, because they serve to fix the scale of phenomena, (in a rather general sense) they may be called *scale constants*. One could argue that it is because they are scale constants that a structural theory can give them. That is the first part of the problem. The second part was quite different: it was to show that the scale constant, say 137, given by the theory was actually the number expressed in measured quantities as $\hbar c/e^2$. Eddington's approach to this, as explained in Chapter 8, was to chase through the conventional equations (Dirac's equation for the hydrogen atom, in this case) to see where the scale constants arose and so to identify them. (It is noteworthy that m_p/m_e resists this 'domestication'.) I would put it this way: Eddington in 1936 sees the second part of the problem as being still a matter for conventional theory.

It is not immediately obvious that such a division of the problem would be unfruitful; but as a matter of fact it proved unworkable – the second part could not be carried through satisfactorily. I shall argue below that there is a good reason for this. In 1936 it was not so obvious, but the suspicion was more acute because of what was perceived, as I said above, of the danger of numerology. It will be useful to illustrate this by later resurgences of the 'danger', the later consequences of which have a moral. I begin with (Lenz 1951) which I can reproduce in full:

The most exact value at present for the ratio of proton to electron mass is 1836.12 ± 0.05. It may be of interest to note that this number coincides with $6\pi^5 = 1836.12$.

The essence of numerology is the identification of certain numbers derived without any theoretical reason known for them. In this situation, the only interest in a numerological result must lie in the closeness of agreement of the numbers. Lenz's result is an outstanding one by this criterion. It should be mentioned here that a natural reaction: 'there must be something in this, the agreement is so close' proves in general to be ill-founded (though as we shall see it may apply here). For out of frequently occurring numbers in physics, like $3, 4, \frac{1}{2}, \pi, e, 10, 137, \ldots$ it is possible to derive a large variety of numbers by products raising to powers and sums. A computer calculation to exhaust all possibilities makes it clear that there will be many close coincidences. The simplicity of Lenz's result gives it only a partial immunity from this criticism.

Some twenty years on, I. J. Good offered a guess for the relation of Lenz's result to Eddington's theory, though one at variance with Eddington's quadratic equation. In (Good 1970) he concentrates on Eddington's phase spaces, which he regards as endowed with a straightforward Euclidean metric. The first space has $R_1 = 10$ dimensions, leaving $I_1 = 6$ over, as we have seen. Similarly, the next one has $R_2 = 136$ dimensions and leaves $I_2 = 120$ over. Good considers a formula for the volume of a unit sphere in these phase spaces. In $2n$ dimensions he quotes the formula $V = \pi^n/n!$.[1] Then he recovers Lenz's result in the form

$$m_{\mathrm{p}}/m_{\mathrm{e}} = I_1 I_2 V_1,$$

since $V_1 = \pi^5/5! = \pi^5/120$, and $I_1 I_2 = 120 \times 6$. He says that this 'begins to give it a geometrical meaning'. Good also derives some other numerological results using the same constants together with an amended form of the Bond factor, which he takes as $1/(136\alpha)$, where $1/\alpha$ is the observed value 137.036.

This particular story comes to an end, for the time being, with a pleasantly ironical twist contained in two short papers by Wyler (Wyler 1969, 1971). In the first paper Wyler proposes a 'geometrical' calculation of the fine-structure constant. This involves carrying out some subtle analytical operations on the so-called conformal group, the largest group of transformations under which Maxwell's equations are unchanged.[2] The operations give rise to two numbers whose ratio is $(9/8\pi^4)(\pi^5/2^4 5!) = 137.0360825 \ldots$. Why should this be the reciprocal of the fine-structure constant? The involvement of Maxwell's equations suggests electromagnetic phenomena, but the idea of a fixed charge e, and Planck's constant $\hbar$ both suggest quantum mechanics, which has not entered the calculation. This is Eddington's second part of the problem again. Wyler's remark on this:

This interpretation of the groups $SO(4,2)$ and $SO(5,2)$ leads to a model of the interaction between the Maxwell field and the field of the elementary particles. This model is defined by the reduction of the representations of $SO(5,2)$ to $SO(4,2)$ . . .

does little to answer the question. It is indeed a remark with very much of an Eddingtonian flavour. The very clever mathematical calculation of Wyler and the simplistic counting of degrees of freedom by Eddington may seem very different. But these aspects do nothing to support or refute their theories; on the identification of numbers they are in the same logical position.

One may also ask whether it is possible for both calculations to be right. They appear to start in very different ways and to continue with wholly different techniques. None the less it would be possible that each was (at least approximately) finding the same thing. This can arise when a complicated mathematical calculation surprises one by connecting numbers which are already known to be connected in some other way. An analogy here would be the surprise that one gets on first being told that

$$\pi^2/6 = 1 + 1/2^2 + 1/3^2 + \ldots$$

Here the surprise can be alleviated by studying one of the ways in which the result can be proved.[3] The difference in the present case is merely that we have no knowledge of any underlying theory connecting the calculations.

In Wyler's second paper he extends the operations carried out and derives a result that simplifies to Lenz's $6\pi^5$. But the difficulties here are even greater than in the first paper, and perhaps than Eddington's. He can remark only: 'The ratio of masses is then obtained by the reduction of Green's functions on Q^3 and Q^4 . . ..'

The importance of RTPE

I turn now to my own views on RTPE and its importance. I have analysed two of its arguments in detail and two more shortly. Of those analysed in detail I have surmised that the first one, on the value of the fine-structure constant, could perhaps be completed to the satisfaction of the community of physicists. Such a completion would, however, be at the expense of an extensive reconstruction of quantum mechanics to take account of the implied environment. Such a completion of the other three is much more in doubt. Bearing this in mind, consider again the realisation which, I mentioned earlier, had begun to dawn on Eddington as he wrote RTPE. It was one thing to devise a structural theory giving rise to certain numbers like 137, which could be expected to be physically important because of the way the theory was rooted in the Clifford algebras. It was quite another to show that these numbers were necessarily related to physical constants such as hc/e^2. The distinction I have drawn above between the first of the four arguments, giving the fine-structure constant and perhaps possible of completion, and the other three may be because hc/e^2 is a constant of a kinematical kind (using the term very generally). It involves no explicit concept of mass or force. Even so, the possibility of rescuing the other three is remote and, though it is also perfectly possible that none of the four can be completed in that way, the great importance of RTPE lies in its drawing attention to the existence of the seven structural constants listed at the end of the last chapter. To put the matter no stronger, RTPE shows that, taking any three of the constants to fix units, it is possible to derive numbers approximating to the others as numbers intrinsically involved in various Clifford algebras. This begins to answer the third of my six questions.

But the value of RTPE lies not only in drawing attention to the existence of scale constants and their possible relations with algebras but also in raising the further question of how to rescue the structural calculations which give rise to such significant numbers, without raising the second stage problems of identification, which seem insoluble. It is hardly likely that a theory giving rise to such a number of numerical agreements can be wholly wrong. Now the only way in which one could justify 137, for example, without showing it to be a *derived* constant, derived that is from h, c and e^2, would be to formulate a theory in which it was not derived but logically prior. It is true that Eddington cannot be credited with making this further step but he reached a situation from which it is not possible to extricate oneself in any other way. Such a step is not without a big price of its own. If the scale constants are prior, it is necessary to show how the appearance of other numbers can be understood. Eddington did not get so far as that problem, but he unconsciously anticipated the way ahead. For the triumphal conclusion of RTPE quoted in the last chapter, though it overstates its case, does so in this way:

There is nothing in the whole system of laws of physics that cannot be deduced unambiguously from *epistemological* considerations.

(My italics.) Now if one considers Eddington's claim carefully, it becomes clear that, if there is anything in it, questions of epistemology cannot be simply pasted afterwards onto a theory like physics which exists independently of them. If Eddington is right in seeing his theory as epistemological, and it is hard to see it in any other way, then this fact needs to be part of the theory from the beginning. The resulting theory cannot be a physical theory in the usual sense, but a theory of how physical knowledge is gained. The process of gaining physical knowledge must be part of the subject matter of the theory. Such a theory has been formulated in recent years by Bastin, Noyes and others.[4] It is beyond the scope of the present book to examine it in detail merely because it arises in response to Eddington's difficulties. Suffice it to say that the theory has no obvious connection with Clifford algebras but it gives, like Eddington's, 137 as a first approximation to a scale constant. Its strength lies in its being able to use this numerical value to identify the scale constant with the fine-structure constant and to carry this identification forward to the finding of the next approximation. This further step gives 137.035 . . . as a second approximation. The theory also gives good reason for the dimensions of space to be three and it generates in a natural way the scale constant $2^{127} = 10^{38}$. To describe these achievements of the new theory must not be seen as denigrating in any way the achievements of RTPE. Rather, it is from the basic idea of the scale constants as prior, with its inevitable consequences, that the new theory is generated, so that it can be seen as a logical development from RTPE.

Notes

1. This is a standard result which can most easily be derived like this: by slicing up an $(n + 1)$-dimensional sphere of unit radius and volume V_{n+1} into n-dimensional spheres of differing radii r and so of volumes $V_n r^n$ one easily gets, by integrating

$$V_{n+1} = 2 \int_0^1 V_n r^n dh, \text{ where } r^2 + h^2 = 1.$$

Put $r = \cos \theta$, $h = \sin \theta$ so that

$$V_{n+1}/V_n = 2 \int_0^{\pi/2} \cos^{n+1}\theta d\theta = \frac{\Gamma(\frac{1}{2})\Gamma(1 + n/2)}{\Gamma(3/2 + n/2)}.$$

Repeating gives

$$V_{n+1} = V_n/[(n + 1)/2] \text{ or } V_{2n} = V_{2(n-1)}/n.$$

Since V_2 is well known to be π, the result follows.

2. Wyler's argument is as follows: M^n is the $(n - 1)$-space-dimensional Minkowski space and the conformal group $C(M^n)$ is the invariance group of the light cones.

It is isomorphic to the group $SO(n,2)$ of the quadratic form of signature $(n,2)$. Then the set $T'' = R'' + iV''$, where V'' is the set of real n-vectors whose first component is positive and dominates the (Euclidean) length of the other components, is the Siegel domain of the hermitean symmetric space $SO(n,2)SO(n,2)/SO(n) \times SO(2)$. It can be realised as a unit disc:

$$D'' = [z \text{ in } C'': 1 + |zz'|^2 - 2\bar{z}\bar{z}' > 0, |zz'| < 1]$$

with z' the transpose of z. Corresponding to a point z_0 on the frontier of D'' there is an analytic map $F: D'' \to T''$ inducing an isomorphism $F_*: I(z_0) \simeq L(T'')$ between the isotropy group of the action of $SO(n,2)$ and the group of linear transformations of T''. Now $L(T'')$ is isomorphic to the Poincaré group $P(M'')$, the semi-direct product of the Lorentz group $SO(1,n-1)$ and the translation group.

If g is any member of $SO(N,2)$ and $j_g(z)$ is the Jacobian of the map $z \to g(z)$, then for every p (an integer greater than zero) one can define a representation of g (the whole constituting a representation of $SO(n,2)$) by

$$R_p(g): f(z) \to f(g^{-1}(z))j_g^{-p}(z).$$

But D'' has a Kählerian structure and the holomorphic $f(z)$ are harmonic so it follows that in the representation the Casimir operator is zero. For the case $n = 4$, this gives the light cone; one gets instead the 'mass-hyperboloid' $E^2 - \mathbf{p}^2 = m^2$ by starting instead with the representations of the group of the Lie geometry of M^4, isomorphic to $SO(5,2)$. This leads him to the 'model' mentioned in the text.

The reduction of the representations needs a little more calculation and a further assumption – that in the reduction the coefficient $e^2/\hbar c$ is given by the ratio of the coefficients of Poisson kernels on D^4, D^5. This assumption is not taken out of the blue; it hinges on the fact that $e^2/\hbar c$ enters as the coefficient of the fundamental solution of the Laplace equation in M^4. None the less, it is an additional assumption. But in Wyler's favour may be mentioned that his value is quoted by him as $137.037\ldots$ whereas it should be $137.03608\ldots$ which is very much nearer to the latest value of $137.0359\ldots$.

3. Perhaps the simplest of these is to evaluate the Fourier expansion of the function sgn x where

sgn $x = 1$ for x between 0 and π, sgn $x = -1$ for x between $-\pi$ and 0,

and for convenience sgn is defined to have period π. Then Fourier's theorem states that sgn $x = \Sigma a_n \sin nx$, where as usual

$$a_n = \frac{1}{\pi} \int_{-\pi}^{\pi} \sin nx \cdot f(x) dx = 4/\pi, \text{ for odd } n.$$

Thus sgn $x = (4/\pi)(\sin x + \frac{1}{3}\sin 3x + \ldots)$ and so

$$\int_0^\pi \text{sgn } x dx = \frac{-4}{\pi}[\cos x + (\cos 3x)/3^2 + \ldots]_0^\pi.$$

Hence

$$\pi = \frac{8}{\pi}\left(1 + \frac{1}{3^2} + \frac{1}{5^2} + \ldots\right)$$

from which the result in the text follows easily.

4. Much of this work is not yet published so a summary is given here. The basic idea is to study the process of increase of information about the world. The model of this chosen is that involved in a division between that part of the world that is known and what remains unknown. Entities may change from being in the unknown part (when nothing can be said about them) into the known part. They then show up as new mathematical elements and the frequency of appearance of such an element is the one source of information about the world. The process is autonomous, so that the requirement of quantum mechanics of 'incorporating the observer' is satisfied without the need to make an untenable distinction between observation and other operations and without the temptation to ascribe properties to human observers.

Each new element has to be labelled say as (a,n) where a is the label of the element and n is an integer stating how many times the element has occurred. The following discussion is concerned with a. It is argued that the method of carrying out labelling is immaterial, so that the process will have the same character as if it were systematically carried out with some fixed label alphabet, L. The particular alphabet $L = [1,2,3,\ldots]$ is adopted for analysis, as it can be without loss of generality. Here the symbols of L are not cardinal numbers but they are used in an ordinal fashion with the obvious ordering. The labels are strings of symbols of L or 'words in L'. The labelling requires a test of whether an element is new or not. In whatever way the process actually operates, the effect is the same as if it operated thus: S is the set of already labelled elements. The process relates S and the not-yet-labelled element; the result is to 'signal' whether the new element is a member of S. The signal is a word in L if the entity is in S. Otherwise it will have a value outside $W(L)$ (the set of words in L) and this value can be taken as one fixed value and written as 0. In all this it is the process that is taking place that is emphasised, rather than the objects taking part in it. Nothing is known about the background (by definition) and so at each stage all possibilities must be treated indifferently; and a consequence of this is that if the process continues long enough any possibility will in due course occur, which is thought of in the theory as a kind of primitive ergodic principle.

An analysis of the special case in which S has only one element, b say, so that the question of whether a is in S is that of whether a is equivalent to b, shows that the requirements of an equivalence relation mean that: (i) Attention can be confined to a subset R of $W(L)$, called the set of rows, of the form:

$$r = r_1 r_2 r_3 \ldots r_k, \, r_1 < r_2 < r_3 < \ldots r_k$$

where the r_i are symbols of L. (ii) There is a map

$$\text{row}: W(L) \to R$$

constructed by removing any pair of occurrences of a symbol of L and reordering the remainder. (iii) The signal generated for $a, f(a,b)$, can be written in terms of an associative and commutative operation $+$, $f(a,b) = a + b$, where the operation $a + b$ is defined in terms of concatenation of rows by:

$$a + b = \text{row}(a \cdot b).$$

Because of the way in which the operation is used, it is called discrimination.

The general case in which S has more elements cannot be treated simply by testing the new element against each in turn of the existing elements of S since such a process would also need to ask, for each element, 'Has this element been tested before or not?' and so on in an infinite regress. The set S has to be treated as a whole. An analysis on the same lines as in the one-element case shows that: (i) an unambiguous labelling by this process arises only if the sets S are *discriminately closed* (that is, such that $a + b$ is in S for any two *different* a,b in S); (ii) there is no loss of generality in taking the signalling process to be defined by a linear characteristic function (because of (i)) S (using the same symbol for set and functional process). The value of $S(a)$ is in R if a is not in S but is 0 otherwise. When such functions S,T are defined, an operation $S + T$ is induced by the rule:

$$(S + T)(x) = S(x) + T(x)$$

for all x in play up to the point reached. (Here $+$ denotes discrimination on the right-hand side.) This operation between S and T also turns out to be discrimination operation but at a 'higher level'. It is therefore possible for the process to ascend to a higher level at which single elements stand for sets of elements at the lower level. It need not do so since the process is self-organising but the ergodic principle shows that eventually it will do so.

There is a limit to the extent of this self-organisation and this limit is given by the construction of Parker-Rhodes: consider any set S of r elements. These generate a discriminately closed set of $2^r - 1 = r^*$ members. Arbitrary discriminations between them will yield one of them or zero, i.e. one of 2^r cases. To specify a member means giving r bits of information, or for shortness, each element carries r bits. The number of discriminately closed subsets generated by subsets of S is also r^* and so, in level changes, any of the corresponding r^* characteristic functions is specified by listing its effect on each member of S–r bits for each of r elements. Thus each function carries r^2 bits. A system of levels can therefore start like this so long as $r^2 \geqslant r^*$, which limits r to 2, 3, 4. To subsume a second level under the previous construction, regard the r^* elements as a subset of r^2 ones each carrying r^2 bits. Then a second level change is possible if $r^4 \geqslant (r^*)^*$ and this limits r to 2 (in which case a further level change is also possible). The bounding construction therefore begins with two elements. At the first stage they define $2^* = 3$ discriminately closed subsets with three corresponding characteristic functions each carrying four bits. These give rise to a further $3^* = 7$ discriminately closed subsets, making ten in all, and each corresponding function carries sixteen bits and gives rise to 127 more discriminately closed subsets, bringing the total to 137. Now these 127 characteristic functions each carry 256 bits but give rise to $127^* = 10^{38}$ more discriminately closed subsets, which terminates the construction because $r^2 = 65\,536 < r^* = 10^{38}$.

The theory argues that the three-fold basic characteristic of this hierarchy of levels corresponds to a three-fold structure of experience (since the process is that of increasing knowledge about the world) and so the bounding construction of Parker-Rhodes demonstrates the three-dimensionality of space. The successive numbers 3, 10, 137, 10^{38} are identified as scale constants in the sense used in the text and the numerical values of them then indicate that the third and fourth correspond to electromagnetism (the fine-structure constant) and gravitation respectively. This suggests an identification of the two others with strong interactions but this seems at present less clear.

Once 137 has been identified as a first approximation to $1/\alpha$, the logical position changes. The particular scale constant is now known and better approximations to it can proceed as follows. Suppose the process is artificially constrained to be operating at the first three levels only and that these three levels have been filled. There are 137 elements and so, in subsequent operations of the process, subject to the constraint, there is a probability 1/137 of any particular element arising again. This interprets the first approximation as a probability in the process. Now if all constraints about level are removed, so that when an element arises it is not determined whether it is at one level rather than another, then at the first level the four possibilities, of its being one of the three elements or being at a higher level, must all be given equal probability $\frac{1}{4}$. Similarly at the next level, 1/8 and at the third 1/128. The probability of being at none of these levels and therefore at the top level is $1/(4 \times 8 \times 128)$. The probability of being any particular element at the first three levels is reduced accordingly and 137 is increased to 137.033. This second approximation is in error by only 0.002%.

The strength of this argument appears, however, when it is realised that it is not quite correct and that correcting it further improves the agreement. The point is that there are 74 088 possible sets of characteristic functions for the seven discriminately closed subsets at level 2 and of these only 61 772 give rise to 127 such subsets at the next level. The remainder give rise to fewer. The factor 1/128 therefore needs to be increased and a lengthy calculation shows that, to a good approximation, it should be replaced by 1/122.229 giving rise to a corrected value of 137.03503, in error by less than 0.001%.

Only imperfect versions of this work have so far been published (Bastin 1966, Bastin *et al.* 1979, Noyes & McGoveran 1989, McGoveran & Noyes 1991, Kilmister 1992). Work is proceeding very actively and the workers have formed an international group, the Alternative Natural Philosophy Association. Further details can be found in the annual Proceedings of the group's meetings (obtainable from Dr F. Abdullah, City University, Northampton Square, London EC1V 0HB). The group is not monolithic. Some members, particularly in North America, have employed promising short cuts to derive a large number of physical constants. For, example, to concentrate on one familiar to Eddington, they derive

$$m_p/m_e = 137\pi/K, \text{ where } K = \frac{3}{14}\frac{4}{5}\left(1 + \frac{2}{7} + \frac{4}{49}\right),$$

which agrees to better than 0.0001%. Such individual agreements could perhaps be dismissed as numerology, but the large number of good agreements is a good defence against this. By contrast, the UK arm of the group is most concerned to clarify principles before embarking on numerical calculations and has also made considerable progress.

12

The last decade

In this chapter I shall describe the last decade of Eddington's life and answer the remaining three questions I raised – those appropriate to FT. This last decade was less happy than his earlier life. Here again the temper of his internal intellectual activity was in tune with the external circumstances. The world outside Cambridge was initially preparing for war and afterwards involved in it. The depression produced by this failure of humankind to act rationally was augmented for Eddington by his own failure to put over his ideas in RTPE. As I argued in the last chapter, this was evidenced less by the uncomprehending reviews than by the lack of follow-up of the ideas. Eddington became less in evidence at the Royal Astronomical Society and in the University as he busied himself at the Observatory in writing what was meant to be the definitive exposition of his work, *Fundamental Theory* (Eddington 1946).

Schrödinger

There are two personal details of his life in these latter years that bear on this. The first of these arose in 1942 when an invitation came from Dublin. In 1940 de Valera, an applied mathematician at heart, had achieved a long-standing ambition to persuade the Dáil to set up an Institute of Advanced Studies, including a School of Theoretical Physics. His move at that time owed much to the availability of Erwin Schrödinger, who had been dismissed from his chair in Graz not very long after the *Anschluss*. The essentially political life of de Valera had left him little time to keep in touch with academic happenings but he had many friends who could advise him, especially Sir Edmund Whittaker. Whittaker had been Astronomer Royal of Ireland from 1906 to 1912 and always retained his affectionate contacts with Dublin. In 1936 C. G. Darwin (he of the '2 scalars, 2 vectors and a six-vector' of an earlier chapter) retired from his chair in Edinburgh. Schrödinger, then living somewhat unhappily as a voluntary exile from Germany in Oxford, was thought of as a possible successor. Whittaker invited Schrödinger to Edinburgh to discuss it but neither seems to have thought it quite right. So Schrödinger went off to his ill-fated stay in Graz (assuring friends who advised him against it that there was no risk of the *Anschluss*) and Max Born was appointed in Edinburgh. But Whittaker's conversation with Schrödinger had convinced him that here was the man for de Valera's hoped-for Institute. It may be, too, that Schrödinger's nominal Roman Catholic faith commended him more in the Dublin of those years. At any rate, Schrödinger became the first director of the School of Theoretical Physics in 1940.

Now Schrödinger, almost alone amongst the more eminent physicists, made repeated attempts to come to some understanding of RTPE. During his Oxford days he had visited Eddington in Cambridge and his review of RTPE in *Nature* was critical but friendly ('a sketch of unusual grandeur'). There were two ways in which RTPE appealed to Schrödinger. The philosophical grandeur of the project fitted his temperament just as well as Eddington's. Schrödinger had wide philosophical interests, mostly in Eastern philosophy, to which he had come through an exhaustive reading of Schopenhauer, although much of his mathematical work was of a very detailed down-to-earth kind. But in addition there was the natural appeal of the way in which Eddington had proclaimed the need to extend the ideas of Schrödinger's wave mechanics to the whole universe. Certainly Schrödinger took RTPE seriously enough to try to get the details straight with Eddington. A particular stumbling block was Eddington's derivation of his so-called 'central formula of unified theory' on p. 271 of RTPE,

$$mc^2 = hc\sqrt{N/R},$$

where *m* is an appropriate standard mass (whether electron, proton or environment particle is not important for the moment). This relation bears a strong resemblance to the one I derived in Chapter 10 but there is a difference and my derivation was carried out in a different way from

Eddington's. Eddington claimed to be deriving the formula from treating the Einstein universe by two methods, though as I explained in Chapter 10, he never quite carries this out. Any attempt to carry out the proof on his lines inevitably finishes with an equation involving N/R^3 or $N^{\frac{1}{3}}/R$ instead of $\sqrt{N}/R$. Milne had earlier tackled this problem with Eddington without success. Schrödinger corresponded with Eddington about it in 1937 and later that year Schrödinger chose, surprisingly, to talk on 'Eddington's theory of the world' at the meeting in Bologna to celebrate Galvani's bi-centenary. The paper was not well received. Further letters between Schrödinger and Eddington did not really help.

It was not then surprising that when Schrödinger (and Heitler, who had joined him by then at the Institute) organised a Summer School in 1942 with several speakers from Ireland and two visiting lecture courses, one of the visitors was Eddington. The other was Dirac, who spoke on quantum electrodynamics, which was still at that time in a very shaky state. For Eddington it was a happy intervention for a fortnight in an increasingly lonely life. No-one from Britain visiting war-time Dublin could fail to be impressed by the difference. Eddington had no great interest in the free availability of food rationed in the United Kingdom, which was the usual preoccupation of visitors to Dublin at that time (or, indeed, for the next ten years). But the general atmosphere of normality must have been important for him. The Institute still occupied the two gracious Georgian houses in Merrion Square where it was founded. Added to all that and to traditional Dublin hospitality a respectful audience was assured for his latest ideas. He had, by 1942, come to terms with the fact that, as a piece of exposition, RTPE had somehow failed. He had not correctly judged the nature of the failure but he saw that he needed to make a fresh start in exposition and so he spoke on 'The combination of relativity theory and quantum theory'. The lectures were published (Eddington 1942). Their content is virtually the same as the first part of FT and I shall discuss the material in that context.

Illness

The second personal detail which bears on Eddington's writing was the state of Eddington's health. This had usually been excellent but throughout 1944 he suffered pain. It was not till November that his doctor moved to an operation. Only his sister was not surprised. He was operated on for cancer on November 7 but without success; he died on 22 November. He had been working very hard all the summer and autumn on FT but he failed to complete it. It is idle to speculate what effect his illness had on the actual composition – whether a clear run of another two years, say, would have produced a clearer book or a more obscure one. But what is not in question is the bibliographical consequences of his death. Whittaker was invited by Miss Eddington and Cambridge University Press to edit the manuscript of the book which Eddington had separated from his other papers, probably on 6 November. He published the work almost unchanged, but he was confronted with the fact that two chapters were unwritten and other sections incomplete. The last chapter of the manuscript is an incomplete Chapter 12; four other sections have headings only. Then Chapter 13 was intended to be an evaluation of the large number N, said to be an expansion of a paper (Eddington 1944) and the last chapter of all, a summary, was to be a section, numbered 137 by Eddington so as to leave us with a jest, on 'The principles of fundamental theory'. Whittaker decided to leave out the missing sections and to print as an appendix to FT the paper on N. The alternative would have been to search through the rest of the remaining material for clues to what would have been put in these chapters. Whittaker probably did not see this as appropriate taking into account the nature of his charge; nor would he had been the right person for the task for his interest was too much in getting the right clear expression of a theory.

The right person was Noel Slater, whom Whittaker had recommended to Miss Eddington to look through the bulk of the manuscripts. Some other manuscripts were found in Miss Eddington's papers in 1954 by A. V. Douglas and F. J. M. Stratton and Slater had access to these in time for the publication of his book in 1957. He had already talked to Eddington in 1942 about the progress of the book and again in the spring of 1944. Eddington had told Slater that the idea of the book was to appeal more to physicists by leaving the algebra to the second part and begin with the basic ideas. His next visit to Cambridge was not till more than six months after Eddington's death. Miss Eddington then showed him a pile of loose sheets, only partly covered. Later G. L. Clark went through these and found nothing 'noteworthy'. These sheets are now lost. There was also an equal-sized pile of chapters, clipped together, which Slater hastily numbered. Two years later Miss Eddington sent most of these to Slater. They were draft chapters of FT which had been abandoned in favour of the final (sixth) version. They were fair copies and Slater concluded that Eddington had preferred to begin a new page rather than delete a sentence, which accounted for the pile of loose sheets. The importance of the draft chapters was not only to see earlier stages in Eddington's thought; it seemed certain that he had referred to them in composing the final version. The result of Slater's painstaking correlation of the drafts with the final book can be seen in (Slater 1957).

Criticisms of FT

The defects of FT are to a large extent those of RTPE and its publication caused a stream of criticism. Some of this was ill informed but all of it is more or less irrelevant to the questions considered here and so I shall not discuss it. But there have been three detailed and more or less sympathetic criticisms of aspects of FT, those of Merleau-Ponty, Dampier and Tupper. These deserve lengthier mention. I shall deal with the first two, which lay bare the true defects of FT, in this section. In succeeding sections I shall discuss Tupper's attempt to reconstruct some of FT and then finally I shall turn to the three questions raised about FT and shall argue that the answers are rather different from those made about RTPE.

As a prelude to Merleau-Ponty's criticisms I shall say something about the principal arguments of FT. The book falls into three parts which are surprisingly distinct. The third part, at first sight, seems not unfamiliar. It consists mainly of a new exposition of the E-number algebra which had started Eddington off on these investigations in 1928. There is, however, a subtle difference, of great importance, which has not been noticed by critics. The algebra is now used almost exclusively in a 'symbolic' manner. The addition operation denotes little more than the conjunction 'and' and the coefficients attached to the Es are labels of qualities rather than numerical quantities. It is a new and highly original way of using an algebra and it is confusing to find that, despite this unusual function, it has now been deposed from the logical primacy which it had in RTPE. The reason, as Eddington told Slater, is mainly one of exposition. The equations determining physical constants in RTPE involved ratios of numbers, like 136/10 and so on. Here the integral numerators and denominators were the dimensions of certain 'phase spaces' using the word in Eddington's somewhat unconventional sense. And the numbers of these dimensions were in turn the number of elements of a certain type in a Clifford algebra. One of the most frequent criticisms that Eddington had heard was the question of *why* such ratios of dimensions should be physically significant. I have argued in the last chapter that Eddington could have answered this only by first moving to the truly radical position that the scale constants are logically prior and second showing how, from that position one could go forward to reconstruct the other results of measurement. No such move was contemplated by Eddington. But he did sketch out a superficially plausible strategy to disarm this criticism. The second part of the book embodies this strategy. It consists in arguing that the requirement of an environment for a physical system studied, together with the usual considerations of special relativity and quantum mechanics, is sufficient to show that ratios of dimensions (of phase-spaces) will give rise to ratios of physical quantities. The simplest example of this is his formula

$$m_1/m_2 = k_2/k_1,$$

(FT, p. 29) for the masses of two particles whose phase spaces have dimensions k_1, k_2. Such a formula must be read with care: m_1, m_2 *may* be actually measured masses but they need not be. It is necessary to investigate in each case how the formula is being used.

This new beginning is not, however, the start of the book. That consists of a short piece of five sections only, which I will deal with in this chapter. This part is not very well integrated with the rest of the book. It may well be that Eddington would have done something about this in a final preparation for publication, had he lived. This first portion is concerned with the 'uncertainty of the origin'. It takes up again the argument sketched in Chapter 10 above, where the physical origin is defined in terms of the centroid of a large number, N, of particles. The reason for this choice (which 'must be' that implicitly made by quantum mechanics) is that the central limit theorem of statistics tells us that such a centroid has a gaussian distribution:

$$F(x_0, y_0, z_0) = (2\pi\sigma^2)^{-3/2} e^{-(x_0^2 + y_0^2 + z_0^2)/2\sigma^2}$$

whatever the (unknown) distributions of the individual particles. This physical origin is assumed throughout the rest of FT. The calculation of the standard deviation of the origin, which gives σ, is carried out in much the same way as I used in Chapter 10 of this book. But it is far from clear how the distribution of particles is envisaged, though in any case, a formula of the form

$$\sigma = AR/\sqrt{N},$$

where A is some number of the order of unity, will result.[1] Eddington argues that this constant, σ, serves to define the unit of length. His reason for this is two-fold. He is to show later in FT that the Rydberg constant for hydrogen involves σ and so a straightforward reason is that taking σ as defining the unit of length fits well with the practice of defining the unit of length in terms of one of the lines of the hydrogen spectrum. But there is a more convoluted argument going on as well. For, as Eddington argues, the definition of units of length and time are required before physics can begin, because of the fundamental character of these quantities. The old alternative – to have a chain of comparisons back to the conventional standard metre in Paris – is inadequate for a fundamental theory, and the definitions in terms of wavelength and frequency of spectral lines is a practically useful device but logically absurd, since ideas of length and time, and much else are needed to observe spectral lines.

This general scheme is enough to understand Merleau-Ponty's position. His criticisms, which are well informed but severe ones, are contained in a book, *Eddington's Philosophy and Theoretical Physics* (Merleau-Ponty 1965). He begins by describing Eddington's work as 'not original but unique' and says 'Twenty years after his death . . . it can be affirmed that the obscurity is, in the end, impenetrable but that it still exercises a power of fascination'. The first part of his book deals with Eddington's work before FT and his discussion of Eddington's philosophy begins, as I have done in this book, with the influence that general relativity had on his thought, as evidenced by his *Mind* article (Eddington 1920b). Here matter is not a substance with real predicates discovered by experiment but an *a priori* mathematical construction from relations. Physical reality 'in itself' is a continuum inaccessible to direct measurement with no ontological consistency outside of the mind conceiving it. Merleau-Ponty contrasts this position with the 'astonishing transposition' in MTR: the methodological perspective displaces the ontological one, the accent is on the 'operational' character of physical concepts. Eddington was on the sidelines in the philosophical ferment over quantum theory. The Einstein–Bohr debates did not affect him; he felt able to adopt indeterminacy with enthusiasm without renouncing the philosophy to which he had been led by general relativity. The enthusiasm was the more because Eddington, almost alone amongst professional scientists and philosophers, though closely in line with lay opinion, always thought that the abandoning of determinism in science was directly connected with the freedom of the will. Having conceived the epistemic unity of the two branches of physics, Eddington claimed to realise this technically. For this purpose, Dirac shows the way;

structure is all and algebra the road that leads to it, though qualified by a reservation given below.

The importance of the idea of a *group*, which captures 'structure', for Eddington lies in the closure property. This is seen as essential as soon as mathematics enters science. If $A,B,\ldots$ are objects, and $P,Q,\ldots$ operations on them, and if $P(A) = X$, then mathematics enters in order to stop the continual multiplication of symbols by claiming that $P(A)$ is another A (or another P). Once Dirac's equation had burst upon the scene, Eddington sees (in the Preface to RTPE) the need for a new philosophy – a redoubling of abstraction and the abandonment of the space-time continuum. The symbolic universe of physics tended to acquire a perfect autonomy and, in a certain sense, became less real in Eddington's eyes.

Some talk, says Merleau-Ponty, of Eddington's 'Pythagoreanism'; a misconception, since numbers are not at all real for Eddington, but there is some justice in the charge, for Eddington renews the approach, perhaps of Pythagoras, certainly of Descartes and Spinoza, of investing a rapidly developing mathematics with an epistemological sense and a fundamental ontology. But this sorts ill with Eddington's views on probability as a measure of our ignorance, which come from his work in astronomy. And, to add confusion, for him probability in quantum mechanics is a frequency, empirical and not epistemic, more objective than subjective.

The exception to the universality of structure noted above is the experience of time. For Eddington immediate temporal intuition is authentic, and this includes the arrow of time. So temporal experience does not belong exclusively to the 'structural' world of physics. The frontier between epistemology and ontology, between structure and matter, passes between space and time – a delicate philosophical position for a relativist.

After this preliminary survey of Eddington's philosophical position, Merleau-Ponty goes on to discuss FT in detail. He lists the difficulties: constantly changing views, changing meaning of terms. 'One never knows for certain from where a formal deduction starts, nor if it succeeds. It is a little like a labyrinth . . .. One believes one has attained an important stage – a place from which large avenues radiate. But it is still a trick of the light, the way out is still distant . . . a fascinating impression, though agonising, of a hidden truth. . ... Is it a work of the imagination, a conceptual game? A work of art, as Milne calls it?' Merleau-Ponty distinguishes three objectives of FT:

(*a*)   derivation of fundamental constants (except for N);
(*b*)   better expression of laws of nature by a suitable algebra;
(*c*)   deduction of N.

There are three methods brought to bear on these:

(a)   statistical, and this is new to FT. It rests on the notion that all measurement is statistical, physical measures are vectors in some vector space and every measure rests on both object and environment;

(b)   algebraic;

(c)   epistemological, or as Kant would say, transcendental.

Broadly speaking, (b) in each case recapitulates RTPE, (a) in each case was dealt with by Eddington in the Dublin lectures and so (c) is the really new part.

In a final section, Merleau-Ponty gives an interesting discussion of three faults of FT. The first one is the fragility of the numerical results. At first sight, he says, one cannot help being struck by their precision and correctness but a closer inspection reveals the subtlety of the methods by which the results are obtained and eventually adjusted! This diminishes their persuasive force. Two important cases exhibit this. Firstly, the fine-structure constant is 137 for Eddington although by 1953 it had been determined as 137.0377 ± 0.0016 (and now would be taken as 137.0359). A 3% error casts severe doubt on Eddington's determination. Secondly, Eddington determines the limiting value of the Hubble constant for nebula recession as $572.36 \, \text{km s}^{-1} \, \text{Mpc}^{-1}$ and the empirical value in his day was about 550. When the better definition of the scale of astronomical distances changed them by a factor of about 7 the effect was to reduce the Hubble constant to 75. It is true that Eddington's value is a limiting one but the discrepancy renders it much less plausible.

Merleau-Ponty then refers to the corrections made (Slater 1947) to Eddington's calculated constants but he concludes that it is not a very satisfactory path to take and he prefers to look at Eddington's 'abuse, transgression and distortion of the norms and procedures admitted by theoretical physics'. He chooses to do this by a detailed comparison between the methods of Eddington and his 'young contemporary, Dirac'. It is clear, he says, that such a comparison must yield valuable information, for Eddington attached great importance to Dirac's work as a starting-point for his own, and yet Dirac has met with universal acclaim and Eddington's theory with discredit.

A glance at the preface of *Quantum Mechanics* (Dirac 1930) confirms the common ground between the two scholars. After he has stated the importance of the theory of transformations, Dirac writes 'This state of affairs is very satisfactory from a philosophical point of view, as implying an increasing recognition of the part played by the observer in himself introducing the regularities that appear in his observations, and a lack of arbitrariness in the ways of nature . . .'. This is evidently very much in the spirit of selective subjectivism. Like Eddington, Dirac sees physical concepts as greatly different from common-sense ones, and he also agrees that although the language of mathematics may be essential in physics, it is only an instrument and one has to try to understand the physical ideas without constant recourse to their analytical expression. Again, Dirac's evident preference for more abstract and algebraic methods, because they seem to penetrate deeper into the nature of things, is a position closely akin to Eddington's structuralism. Moreover, it is not simply a greater degree of mathematical rigour which gives Dirac the advantage, for his introduction of the delta-function was at the time quite unjustified, and seen to be so.

The difference is to be found rather, argues Merleau-Ponty, in the style and movement of their thought. To the labyrinthine character of FT is opposed the simple order and regular developments of Dirac's book. First come the physical principles, then the formal development of the algebra, followed by more and more complex applications. Finally, neatly separated, comes the extension of the theory to special relativity. One never finds, as with Eddington, an equation given a different sense from that of its discoverer, or from tradition. This has to be contrasted with Eddington's 'abuse of power'.

In his final judgment, Merleau-Ponty points to the fact that no significant errors have been found in Eddington's mathematics and concludes that "one cannot attribute the obscurity of Eddington's later writings to any enfeeblement of his mathematical powers, nor any 'going off the rails' of his logic". It must rather be attributed to the 'abuse of power' by Eddington. 'He takes bizarre concepts invented by others (for example, exchange forces) and uses the term for a different concept or in a figurative way. He abuses the fluidity of description which physics allows to permit himself a multiplicity of perspectives.'

Dampier's analysis of FT is contained in his thesis (Dampier 1969) which is mainly concerned with the algebraic structure. But he also makes a number of useful general points on which I concentrate here. He notes that Eddington's basic idea is that general relativity and quantum mechanics are different approximations to the world and so are not to be 'reconciled'. Instead, Eddington's intention is to seek the small area of overlap in order to find connecting links, and these links will be evidenced in numerical constants. Such a programme leads one to imagine a situation in which there is some Ur-theory, as it were, which can become general relativity or quantum mechanics respectively, by means of certain limitations. This situation is hardly to be found in FT. One might look, for example, at the appendix to FT, since this begins with a logical analysis of measurement. But this is isolated from the rest of FT and so no Ur-theory is to be found there. Or one might look at the earlier chapters but again the medley of subjects provides no trace of an Ur-theory.

Dampier notes that the calculation of constants in FT is on a much wider scale than in RTPE. The ones calculated fall into three groups. In §32 of FT are listed fifteen 'atomic' constants, in §52 a further twelve 'molar and nuclear' ones and further nuclear constants are to be found scattered in Chapters 9–12. Now, argues Dampier, one might well expect some constants to be determined. Apart from the arguments of RTPE, Chapter 9 of *New Pathways in Science* (Eddington 1935) argues that any true unification of theories will tend to eliminate constants. Moreover, constants represent the limits of explanation of theories, in the sense that, for example, in gravitational theories, whether Newton's or Einstein's, there is no explanation of the constant of gravitation G. But the best analogy to pursue is the elimination of the speed of light by electromagnetic theory. Eddington carries this part of his programme to an extreme. He contemplates reducing the number of uncalculated constants to one large one, N, the 'cosmical number'. Surprising though this would be, in principle it is not unorthodox. In the appendix to FT the programme is taken one step further; N itself is calculated. This is a radically new departure and if it is to be accepted, our notion of physics has to change completely.

But, argues Dampier, consider now Eddington's actual achievements rather than his programme. Since the basic constants are dimensionless one must insert values of three of them to fix the units of measurement. Eddington chose c, the velocity of light, R, the Rydberg constant for hydrogen because it is very accurately known from spectroscopy and F', the Faraday constant for hydrogen. It should be mentioned at this point that Eddington uses the Bond factor, $\beta = 137/136$ widely, to various powers. But this use is subsidiary and can be ignored for the present.

His table I lists 27 constants:

1. M mass of the hydrogen atom
2. m_0 mass of the comparison particle
3. μ mass of 'intracule'

4.	m_e	mass of electron
5.	m_p	mass of proton
6.	e	elementary charge
7.	e'	the same, 'molarly controlled'
8.	h	Planck's constant
9.	h'	the same, molarly controlled
10.	h/e	direct method
11.	$e/m_e c$	direct method
12.	e/Mc	Faraday constant, F
13.	e'/Mc	Faraday constant, F'
14.	m_p/m_e	mass-ratio
15.	$hc/2\pi e^2$	fine-structure constant
16.	G	constant of gravitation
17.	N	particles in the universe
18.	R_0	Einstein radius of space
19.	R_0	the same (in megaparsecs)
20.	M_0	mass of the universe
21.	ρ_0	density of Einstein universe
22.	V_0	nebular recession (km sec^{-1} mp^{-1})
23.	$e^2/Gm_p m_e$	force constant
24.	k	nuclear range constant
25.	A	nuclear energy constant
26.	$A/m_e c^2$	direct method
27.	σ	uncertainty constant

However, if one ignores the Bond factor by taking it as unity, there are three duplicates in this list (6,7), (8,9), (12,13) and 12 is also one of the defining constants and so is not calculated. Proceeding through the list in order, we can take 1 as an independent constant, but then this determines 6. The four 2, 3, 4, 5 are really only two independent constants. From these and R, 8 is determined and so 10, 11, 14, 15, 16, 17, 18 are new constants as is 20, but then these determine 21, 22, 23. Finally, 24 is new as is 25; these determine 26 but 27 is new. The list is reduced to at most nine constants, taking account of 1 being the sum of 4 and 5.

Dampier analyses the list a little differently and claims:

We shall show that all the determinations of the atomic constants depend on these two calculated values.

(He means by this, the values of 14, 15, the mass-ratio and the fine-structure constant.) His proof is on these lines: One can begin with some mass-ratio, say $M/\mu = (m_p + m_e)^2/m_p m_e = A$ say. Then one can also take $\hbar/e^2 = B$. To determine cgs units one takes

$R = (\text{const.})\mu c/2\pi\hbar\alpha^2$, giving $\mu = Ch$,
$F' = e'/Mc$,   giving $e = DM$,

where A, B, C, D are some known constants whose values do not concern us. But from these equations one easily sees that $MABCD^2 = 1$ and this determination of M leads to others of $\mu, \hbar, e$. Then in the second part of the table we have N and Eddington's infamous 'central formula of unified theory'

$$m_0 = E\sqrt{N}/R_0$$

for some constant E. The range constant σ is defined as $R_0/\sqrt{N}$ and so can be discounted. R_0, M_0, ρ_0 are related by general relativity and this is always taken for granted by Eddington; whilst the Hubble constant V_0 is determined in the specific expanding universe model which he also takes for granted. In this way R_0, G and ρ_0 are successively determined and so the whole table is found, except for three nuclear constants, from just four calculations:

$$m_p/m_e, \ \hbar c/e^2, \ m_0 \ \text{and} \ N.$$

If the idea of a calculation of N is disallowed, it is still the case that the other three are to be found in terms of N.

The upshot of Dampier's analysis as a whole is considerably less unfavourable than Merleau-Ponty's, but as far as the calculation of constants is concerned the advance from RTPE to FT is shown to be an illusion. The new constants – Eddington's claim to substantiate his theory notwithstanding – are merely determined from orthodox formulae in terms of the old. Dampier follows this severe criticism with a more positive view of Eddington's achievement. He begins this by asking a question about the nature of physical theories. He defines: 'Theories that take some account of the process of measurement we shall call fundamental physical theories.' The two examples that spring to mind are special relativity and quantum mechanics. In such theories one expects part of physics to be the formalisation of the logical structure of measurement. Dampier formulates a weak form of 'Eddington's Hypothesis' to explain how this would show up; it is evidenced in 'some kind of restriction upon the formal system of such theories'. The task is to find the restrictions explicitly. His conclusion is

We do not claim it is proven, but Eddington has done a service to physics by raising the question of epistemological physics. We can hardly rest content until we have so completely examined this approach that we can either carry it to a successful conclusion or definitely prove that it is untenable.

The critical study of Tupper: scale

Tupper's study (Tupper 1957, described in Kilmister and Tupper 1962) avowedly concentrates on the 'statistical theory', that is, the part of FT derived from the Dublin lectures which is different from and only dimly foreshadowed in RTPE. But he also sets the statistical theory in its context and his conclusions can be most simply set out in the form of his 'extremely simplified scheme' which connects together in a logical order some parts of FT which he finds lend themselves to it. By so doing, he reconstructs these parts of FT into a more acceptable form. He concentrates on four aspects of FT:

(i)   Eddington's notion of *scale* and its relation to the idea of the *environment* of a physical system, a relation which was taken for granted but not investigated by Eddington.
(ii) Eddington's notions about the 'non-Coulombian potential' and the potential for nuclear forces.
(iii) Eddington's division of scale-free and scale-fixed properties and his arguments from it about the form of the energy tensor.
(iv) The derivation of the proton–electron mass-ratio.

It will be sufficient to give an idea of Tupper's approach and its success if I explain only the first two of these in detail.

Tupper begins with an analysis not to be found in Eddington of the concept of *environment*. He does this in terms of an approach which regards scientific theories as abstract structures which may have different degrees of complexity, when dealing with the same subject matter. In that case, there will be experimental procedures appropriate to each degree of complexity; it would be wrong to use experiments in a more complex structure to 'discredit' the theoretical results in a simpler one. The notion of environment is then made explicit as that of a complex abstract structure and a simpler one 'contained within it'. A special case of this containment has been explored already in this book (Chapter 8, especially the discussion of idempotent *E*-numbers). Eddington's explanation is more picturesque; he envisages a description of a determination of a mass. Usually the experimenter omits any reference to the standard kilogram housed in Paris. Eddington asks how this can be. A complete description must include such a reference; but not of course in all its possible detail, which would be counter-productive. Rather, something simple which would still perform the duty of the standard is required. This leads him to the notion of comparison particle (as in Chapter 8 of this book) or, equivalently, to the notion of the 'perfect object system' which includes both object and comparison particles. In the perfect object system the comparison particle is simplified to just one variable, called the *scale*. The analysis of this, though not to be found in Eddington's work, leads directly to the importance of Clifford algebras and so forges a link which he was unable to make between the two parts of FT. The basic idea is that of an observable of order n which is a set of n numbers, one of which (if this is a perfect object system) must be the scale. Then a comparison of two observables, s,t say, will have the form of a linear relation

$$t_u = P_{uv}s_v$$

(with the summation convention). Since P is a comparison of two observables, it is an observable but of order n^2 and this leads to further observables of order n^4 and so on. The whole set of comparisons of observables of order n will then be contained in the algebra of $n \times n$ matrices, M_n. In this algebra the product PQ is then defined as the result of doing the comparisons in succession. I leave the technical details to a note[2] but I need to explain some of the extra assumptions that arise in it since in the note they arise simply as simplifications of the mathematics. Although the idea of an observable was presented simply as a set of n numbers, transformations which preserve addition and multiplication will be allowed and the effect of these will be to mix together the various numbers including the scale. The requirement that the scale is to be included has then to be put in the form that a number can be constructed in a linear way from the components which gives the scale. If the observable P belongs to an algebra of matrices, the scale has the value $\mathrm{tr}(BP)$ in terms of the trace operation (sum of diagonal elements) tr, where B is one of the matrices. It may well be that the possible observables P are not all of the matrices of a certain size but form a 'sub-algebra'. In this case B must be allowed, none the less, to belong to the full algebra of all $n \times n$ matrices. The requirement of invariance in such a way of constructing the value of the scale is assumed to restrict the possibility of transformations but not to disallow them

completely ('first scale condition') and such a restricted set of transformations is assumed to characterise its scale operation ('second scale
condition'). It is then possible to show that, without any loss of generality,
one can take $B^2 = \pm 1$ or 0 and it is assumed that the further case $B^2 = 0$
can be disregarded. The final assumption is that the set of all possible
transforms of B ('the scale-algebra of B') is equivalent to the set of
observables. In such a case, the scale algebra is proved to be a Clifford
algebra and any element of its basis can be taken as a scale operator. The
exhibiting of this clear mathematical connexion between the notion of scale
from Part 2 of FT and that of Clifford algebra from Part 3 is a major
achievement of Tupper's work and goes some way to showing an internal
consistency in Eddington's thought which formerly appeared to be lacking.

The critical study of Tupper: non-Coulombian potential

Tupper now follows Eddington in stepping back from the result above
which he has proved and which Eddington assumed. The scale, they both
argue, is just one way of including the environment. But the notion of
environment is also connected with that of a coordinate-system (consider,
for example, how the presence of curvature in general relativity prevents the
adoption of a universal cartesian coordinate-system). So the next question
is of how to relate the coordinate-system to the environment and so to the
scale as above. The first answer to this is provided by the derivation of the
formula

$$\sigma = R/3.1\sqrt{N}$$

earlier in this chapter. The version given there is different from Tupper's; I
believe that, despite the care with which Tupper corrected Eddington,
mistakes still survive in his treatment of what is inevitably a very subtle
calculation.

The second stage, with Tupper as with Eddington, is to use this result to
tackle the problem, as Eddington saw it, of the nuclear potential. Very little
was understood about the forces in the nucleus in Eddington's lifetime and
less still in his own body of knowledge. However, if one considers the usual
potential function $1/r$, where r is measured from the origin, it is clear that
there will be a difference according to whether r is measured from the
mathematical or physical origin. Such a difference is quite negligible unless
r is very small – of the order of nuclear dimensions. Eddington seizes on this

to tackle what was at the time one of the least understood parts of quantum mechanics, the theory of the nucleus. He concludes rather hastily that the 'nuclear potential' must have a gaussian form, Ae^{-r^2/r_0^2}. Two points must be made about this. Firstly, it was the case that such a potential was used in the late 1930s and early 1940s in nuclear physics. But it was in the context of an approximate theory of a pragmatic kind which was quite at variance with Eddington's approach. But secondly, Eddington's assumptions do not in fact yield this result[3] but rather a field of the form $2\mathrm{Erf}(r/r_0)/r$, where the error function is defined by

$$\mathrm{Erf}\, z = \frac{1}{\sqrt{\pi}} \int_0^z e^{-x^2} \mathrm{d}x.$$

Eddington goes on to say (FT, p. 10) 'we need not hesitate to reject the meson-field hypothesis altogether'. Here he refers to Yukawa's theory (see Chapter 9 above) that nuclear forces result from the exchange of a massive particle, known as a meson. It is rather as when two skaters are moving along converging lines and one throws a heavy ball which is caught by the other. The paths will change because of the momentum carried by the ball. Yukawa's theory predicts a potential of the different form $e^{-r/r_0}/r$.

The two different forms do not present quite the absolute contrast that Eddington implies. The 'nuclear potential' is really regarded as an addition to the ordinary potential arising from the Coulomb field and so the question is whether the result that Eddington's suggestions find, when corrected as in Note 3, do represent the difference between a Coulomb and a Yukawa potential. A simple calculation[4] shows that the difference is well approximated by a Yukawa potential, so that Eddington's rejection, if it is empirically based, is unsafe.

It is a commonplace that the determination of the parameter of a Yukawa potential is a very inaccurate process, so it is hard to get an exact value. However, the argument of Note 4 leads to a most likely value of $r_0 = 1.773\sigma$. Tupper, following Eddington, goes on to compare this potential with that which describes pp-scattering, at least at fairly low energies. Taking the expression above for σ in terms of R and N and using $V_0 = 1/R\sqrt{3}$ for the limiting value of the Hubble constant we have in round terms (and in cgs units)

$$V_0 r_0 = 1.773c/3.1\sqrt{N},$$

which, with Eddington's value for N and putting V_0 in km s^{-1} Mpc^{-1} and

r_0 in units of 10^{-13} cm, gives $V_0 r_0 = 150$. Now V_0 is somewhat less than 100 so that r_0 is of the order of 2. The experimental value of $r_0 = 2.1$ shows that this rough calculation is on the right lines. However, there is a serious defect in this analysis; it applies only to charged particles. It would be useful for pn-scattering only if the neutron were analysed in a suitable way. Even such an analysis will not save the day for, because of the dependence on charge, the strength of the alleged nuclear field is too small. It has the value e^2/r_0, which is about 0.8 MeV and so in error by about 30:1. Tupper concludes, in line with Eddington, that this analysis applies to notional 'standard particles' and that these standard particles are not protons. He gives a somewhat tentative argument for this putative nuclear field to be proportional to mass, which would suggest a standard particle of approximate mass $1836/30 = 61.2$ electron masses. This is not an unhopeful guess, for $68.5 = 137/2$ has often been noticed as a natural mass unit for the less unstable particles.

Tupper's criticism: Summary

I have dealt in detail with two of the four major reconstructions that Tupper has made of FT. The two remaining investigations, added to these, finish with four equations which I give here in a corrected form:

$$\sigma = R/3.1\sqrt{N}, \tag{1}$$

$$r_0 = 1.773\sigma, \tag{2}$$

$$M = 2mN, \tag{3}$$

$$\sigma = \hbar/2m. \tag{4}$$

Here M is the mass of the universe, which Eddington always assumes (conservation of mass) to be that of the initial Einstein universe, M_0, given by

$$M_0 = \pi R_0/2G, \tag{5}$$

where R_0 is the initial Einstein radius. Also m is some standard mass in terms of which the universe is analysed. The argument for (3) hinges, according to Tupper, on the factor 2, for otherwise it simply expresses the analysis of the total mass of the universe into N equal particles. The factor 2 is held to arise because the standard particles are charged and so two are needed to make a neutral particle. Tupper uses much the same argument to account for the factor 2 in (4), which is, otherwise, merely a recognition that σ is expressible as a Compton wavelength (of *some* particle, but in the very simple system considered by Tupper, this must also be the standard particle).

To extract something from these results, Tupper has recourse to some cosmological results. He assumes with Eddington the standard expanding universe, puts $R = aR_0$ to denote the state of expansion reached, where a is in the range 1–10 and uses the standard formula for the Hubble constant:

$$V = \frac{1}{\sqrt{3R_0}} \sqrt{\left[\frac{(a-1)^2(a+2)}{a^3}\right]} = \frac{f(a)}{\sqrt{3R_0}}. \tag{6}$$

He then eliminates the unmeasured quantities σ, R, N, m, M and (converting the results to cgs units) derives

$$1 = \frac{(1.773)^3 2\hbar G T a^2 f(a)}{\sqrt{3} \times (3.1)^2 \pi r_0^3 c^2}$$

(writing T for the reciprocal of V). By assuming a Hubble constant of $75\,\mathrm{km\,s^{-1}\,Mpc^{-1}}$ he finds the right-hand side to be $0.732a^2 f(a)$, so that

$$a^2 f(a) = 1.37$$

corresponding to a value of a of a little less than 1.6. There is, to my mind, some doubt about the factor 2 in (4) and if it is omitted then $a^2 f(a) = 2.74$ making a a little less than 2. Tupper argues correctly that this is a striking confirmation of his work, since it could well have been the case that the answer for a could have been wrong by a factor of 10^{10}. He claims with justice that his arguments have very much less difficulty and obscurity than Eddington's and he believes that a more detailed investigation would remove the remaining difficulties. It is fair to remark that in the intervening thirty years no such investigation has taken place.

The relevance of Tupper's reconstruction is slightly oblique. He points out he has not constructed an *a priori* theory of the kind that Eddington said he was constructing: 'It is not unique, there have been numerous arbitrary choices in its construction, all of which have been made to the best of our ability with one eye on the experimental results.' But the importance of Tupper's reconstruction is that it has in common with Eddington's theories the difficulty of taking place in (at least) two abstract structures of differing degrees of complexity. The theoretical difficulty which arises in doing this is that of comparing results in the two structures. The example of Newtonian gravitation and general relativity exhibits this; the constant of gravitation, G, is the same in both theories but the number of degrees of freedom of the gravitational potential in one theory is 1 and in the other 10. In going from one system to a less complex one, some numbers will be invariant and some will not. Tupper believes that the solution of the problem of determining the numerical invariants of the transformation between such abstract systems will really contribute to the understanding of Eddington's work.

The value of FT

It is time to draw our long investigation to an end. The trail has been a roundabout one, taking in a surprisingly large part of the history of the first half of the twentieth century. I have made no secret of my own opinion of the great value of RTPE and FT although in the nature of things my detailed comments have been mainly on the adverse side. This has been the more so about FT. In this concluding section I turn to the three questions about FT corresponding to those I posed and answered about RTPE. It will be recalled that the questions in the earlier case were, in sum:

1. What made Eddington write RTPE? The answer is that his excitement at the new developments opened up by Dirac's equation for the electron, the apathy and neglect of the scientific community as far as his own papers on this were concerned and the need, augmented by external political developments, to get some clear statement out made him hurry with the publication.

2. Why is RTPE obscure? The book was a workshop not a showroom. The mathematics is rarely above reproach. Added to this, which was unexpected after the beauty of MTR, Eddington's highly unorthodox views on philosophy of science, which had been ignored in the earlier book play an increasing role.

3. What important and valuable aspects does it have? The most important aspect is the working out of an unusual set of philosophical ideas. These ideas may lead to obscurity but they also allow the construction of a theory which purports to connect general relativity and quantum mechanics. Amongst the results of such a connection are numerical values for four constants which are usually empirically determined. The notion that such a Kantian theory, in which a careful analysis of

the method of physics could lead to results that are usually regarded as empirical, may be called *Eddington's conjecture*. This conjecture has a great importance, notwithstanding the imperfections of the attempt that the book makes to realise such a theory.

Apart from that there are at least three valuable points of detail. The first of these is the emphasis on the cosmological side of the argument, which is a reminder that Mach's principle is still an outstanding problem for physics, and the related insistence that some form of environment is needed for quantum systems. The second is the use of the Bond factor, unfortunate though this is. It is certainly used by Eddington to correct results with little clear justification. But the importance lies, not in the Bond factor itself, but in the realisation that such a theory needs to have further development so that a second (and higher) approximation can be found. Thirdly, the emphasis on N again shows that the whole universe is involved.

When we return to FT, the answers are somewhat different:

1. Why did Eddington write FT? The book was written more in sadness than excitement, as a final definitive statement of his views. It was intended to answer the adverse criticism of RTPE. This criticism was mainly that RTPE was obscure; to a lesser degree, it was that RTPE calculated constants, which 'must be wrong'. As I have urged before, this is strange. Not only is theoretical physics primarily about calculating numbers; the oddness of the scale constants had been noticed already at the turn of the century (Planck 1899) in connection with the need for a new one, h, in considering Wien's formula, and the possibility:

... gegeben ist, Einheiten für Länge, Masse, Zeit und Temperatur aufzustellen, welche unabhängig von speziellen Körpen oder Substanzen, ihre Bedeutung für alle Zeiten und für alle, auch ausserirdische und aussermenschliche Culturen notwendig behalten und welche daher als 'naturliche Masseinheiten' bezeichnet werden.
(... of setting up units for length, mass, time and temperature which will necessarily retain their significance independently of special bodies or substances, for all times and even for other worlds and non-human cultures, and which might accordingly be called 'natural units'.)

To understand properly such strange scale constants would surely involve also being able to calculate them. It is true that this view of Planck's has been criticised (Bridgman 1922) because, it was held, Planck could not show any essential connexion between the theories leading to the various constants, but this comes from an extreme positivist standpoint.

At any rate, Eddington concentrated on this lesser criticism and sought to subvert it by calculating many more constants. It is sad that the outstanding intelligence that produced the lucid MTR and the fireworks of RTPE did not see the need to address the obscurity directly. Instead of clarifying the earlier ideas he launched many new ones.

2. Why is it obscure? The book is unfinished, a fact which cannot be entirely ignored. But it seems likely that, had he lived, Eddington would have done nothing to decrease the obscurity. The fact is that FT has an obscurity different in kind from RTPE. I believe that a major part in the causing of this was played by Eddington's truly great achievements in popularising science. He achieved his stated ambition of writing not one, but several, hugely successful science books. One has to ask, what is the nature of such an achievement? At the risk of oversimplifying, one could say that it consists of telling less than the whole truth (for telling the whole, if it were possible, would be to address the scientific community) and of doing so in such a way that the absence of the whole is not obtrusive. This gentle sleight of hand is accomplished by the use of apt analogies and pictures. There is no place for such techniques in a book like FT but whether because, after his successes, it had become second nature to Eddington, or because he saw no other way of skating over some very thin ice, the book suffers greatly from them. The lucid and eloquent English style, unchanged since MTR, is now pressed into service to conceal the gaps in argument or inconsistencies. The other reasons for the obscurity are much the same as in RTPE, mathematical fudges and a highly unorthodox (but largely unstated) set of philosophical views. But the wealth of new physical ideas, no more clearly stated than those in RTPE adds to it.

3. What important and valuable aspects does it have? The virtues of FT are no more than those of RTPE. The recognition of the importance of scale and the emphasis on the need to consider the environment of every physical system are put more strongly. But the position about the calculation of constants has not changed.

Of the two books, FT has had the greater readership and this is because people have tended to accept Eddington's own view that it is the authoritative final statement on what was only in draft ten years before. R. B. Braithwaite's warm but critical notice in *Mind* (Braithwaite 1940) of *Philosophy of Physical Science* concludes:

Since Eddington now regards himself as investigating primarily the nature of physical knowledge rather than the nature of the physical world, he no longer talks of different contrasting worlds (of scientific tables and of familiar tables) in the way which (in *The Nature of the Physical World*) Professor Stebbing and other philosophers have found so confusing. In the present book it is the substance and not the mode of expression of his doctrine that will shock empiricists. Many 'epistemological empiricists' from Mach onwards have called attention to the fact that sentences in physical treatises which in the past have expressed empirical propositions are now used to express definitions or logical consequences of definitions and that as physics advances there are changes in the conventions determining which sentences express the *a priori* and which the empirical. But, to my knowledge, Eddington is the first writer to maintain the position, qualitatively and not merely quantitatively different from the proper recognition of the conventional element in modern physics, that every genuinely fundamental law of physics is non-empirical and *a priori*. I cannot close this very critical notice without expressing my admiration for his solitary heroism in defending such a solitary position. What a philosopher regrets is that Eddington has not thought fit to argue his central thesis in the systematic manner it requires. He has yet to give us his *Critique of Physical Reason*.

The more generous of the views of the philosophic community, which Braithwaite sums up here, could well be applied with equal force to FT. Moreover, the view of Eddington as solitary hero had been current for some time amongst intellectuals more generally. To take one example, Walter Benjamin, writing on the basis of a letter he had already written in 1938, says (Benjamin 1968):

Kafka's work is an ellipse with foci that are far apart and are determined, on the one hand, by mystical experience (in particular the experience of tradition) and, on the other, by the experience of the modern big-city dweller . . ... When I refer to the modern big-city dweller, I am speaking also of the contemporary of today's physicists. If one reads the following passage from Eddington's *Nature of the Physical World*, one can virtually hear Kafka speak:

> I am standing on the threshhold about to enter a room. It is a complicated business. In the first place I must shove against an atmosphere pressing with a force of fourteen pounds on every square inch of my body. I must make sure of landing on a plank travelling at twenty miles a second round the sun . . . Verily it is easier for a camel to pass through the eye of a needle than for a scientific man to pass through a door. And whether the door be barn door or church door it might be wiser that he should consent to be an ordinary man and walk in rather than wait till all the difficulties involved in a really scientific ingress are resolved.

In all of literature I know of no passage which has the Kafka stamp to the same extent . . ... What is actually and in a very literal sense wildly incredible in Kafka is that this most recent world of experience was conveyed to him precisely by this mystical tradition.

Here again the same remarks could have been made with equal force about FT.

To my mind, the emphasis on Eddington's work later than RTPE is mistaken. The earlier work leading to RTPE has a simplicity that allows one to get much closer to Eddington's actual thought whereas the later contains much worthless elaboration. For all its faults, RTPE remains, because of its wealth of revolutionary ideas, one of the most important scientific books of the first half of the twentieth century. It puts at the head of its statement of purpose the outstanding problem in physics – the establishment of a sensible relation between general relativity and quantum mechanics. An enormous amount of work on somewhat more orthodox lines than Eddington's towards quantum gravity in *some* sense of the words has still left this problem unsolved. This suggests that perhaps the wrong question is being asked. If that is so, then Eddington's project to relate them rather than to unite them deserves more consideration. But even this is less significant than the way in which the book sets out Eddington's conjecture: that a careful analysis of physical method can lead to some results which are at present thought of as experimental. Such an exciting idea deserves far more consideration than it has had and when it has been satisfactorily carried out, Eddington will be recognised as its principal progenitor. When that time comes, FT will I fancy, be seen as a mere appendix to RTPE.

Notes

1. Eddington is concerned with the properties of a set of particles in the Einstein universe. Consider the case of two particles first. The Einstein universe is isotropic and homogeneous, so one can take one of the particles (assuming that its mass is small enough not to disturb the background metric) at the origin of a coordinate-system in which the metric is

$$ds^2 = dt^2 - \left[\frac{dr^2}{1 - r^2/R^2} + r^2(d\theta^2 + \sin^2\theta d\phi^2) \right].$$

The first question to be settled but neglected by Eddington is: what is the *measured* distance of the other particle which we can take, without loss of generality as at $(r, \pi/2, 0)$? Now $ds^2 = dt^2 - dq^2$ where

$$dq = \frac{dr}{\sqrt{(1 - r^2/R^2)}}$$

or, if one writes $r = R \sin u$, $dq = R du$ and $q = Ru$.

The second question is then: what is the average distance between the particles? One must assume some distribution; and surely the only one in play is: uniformly distributed per unit proper volume. For the spatial geometry alone $\sqrt{g} = R^2 \sin^2 u \sin\theta / \cos u$, easily, so that the mean value of q is

$$\langle q \rangle = R \langle u \rangle = R \frac{\iiint u \sqrt{g} \, dr d\theta d\phi}{\iiint \sqrt{g} \, dr d\theta d\phi}$$

and this easily gives $\langle q \rangle = (2R/\pi)(1.7337) = 1.1037R$. (This may seem large for a mean value, but it is to be expected from the two facts that the greatest measured length is $1.57R$ and that the particles, from isotropy, are more numerous at the greater distances.)

Now Eddington wants a property of a sample of N particles, so he falls back on his extensive knowledge of statistics and considers the standard deviation of the distribution, for then he knows that the corresponding figure for N particles is got by dividing by $\sqrt{N}$. The second moment about the mean is then easily derived as $R^2(\pi^2/12 + \frac{1}{2} - 1.2182) = 0.1043R^2$ and so the standard deviation is $\sigma = 0.3229R$ or $R/3.0969$. This correct version contrasts with Eddington's $R/2$ but agrees very closely with Slater's $R/3$.

2. Allowable transformations are evidently automorphisms and so, for a total matrix algebra are of the form

$$G : P \to P' = qPq^{-1},$$

where q is any matrix of the size assumed, although P may be confined to a sub-algebra of the total. This is the 'first measurement condition'. Such an automorphism mixes up all the components, including the scale, so that the scale assumption has to be rephrased as:

There is a linear functional, σ, of the algebra onto the reals, invariant under automorphisms, called the *scale*. Thus σ defines a subgroup $G^{(\sigma)}$ of G under which σ is invariant and the first scale condition is that $G^{(\sigma)}$ is neither zero nor the whole of G. But of course there may well be linear maps, a_i say, invariant under the whole of G and so, if σ satisfies the above scale condition, so also does $\sigma' = c_0\sigma + c_1a_1 + c_2a_2 + \ldots$. One then refers to σ, σ' as equivalent scales. The second scale condition is that the scale group $G^{(\sigma)}$ determines its scale, up to this equivalence. (This need not be the case, but it is argued by Tupper that it is the simplest theory, to begin with.)

One can evidently write $\sigma(P) = \text{tr}(BP)$ for some suitable B in the total algebra and it is then clear that $G^{(\sigma)}$ is the commutator algebra of B. But if B commutes with q so also does B^2 and so the scale conditions imply that

$$B^2 = c_0B + c_1.$$

Without loss of generality, one can normalise B so that $B^2 = \pm 1$ or 0. Then arguments are adduced to disregard the zero case as singular and so $B^2 = \pm 1$. It is a small step from here to establishing the scale algebra as a Clifford algebra, with the basis elements as the possible Bs and a little more attention to detail is enough to show that it must be an *even* Clifford algebra.

3. One could put $\mathbf{r} = \mathbf{u} + \mathbf{r}'$, where $\mathbf{r}$ is the distance from the coordinate origin and $\mathbf{r}'$ from the physical origin. Then the mean value of the Coulomb potential will be

$$\phi = \left\langle \frac{1}{r'} \right\rangle = \frac{\int \frac{1}{r'} e^{-u^2/4\sigma^2} d^3u}{\int e^{-u^2/4\sigma^2} d^3u}.$$

Write this as N/D, take $\mathbf{r} = re_3$, so that $r' = |\mathbf{r} - \mathbf{u}| = \sqrt{(r^2 + u^2 - 2ru\cos\theta)}$ and $d^3u = u^2\sin\theta\,du\,d\theta\,d\phi$. Moreover, $D = 8(\pi\sigma^2)^{3/2}$ by a standard calculation. Now

$$N = 2\pi \int_0^\infty \int_{-1}^1 \frac{dc}{\sqrt{(r^2 + u^2 - 2ruc)}} u^2 e^{-u^2/4\sigma^2} du$$

$$= 2\pi \int_0^\infty [-\sqrt{(r^2 + u^2 - 2ruc)}]_{-1}^1 \frac{1}{ru} u^2 e^{-u^2/4\sigma^2} du$$

$$= 2\pi \int_0^\infty \frac{r + u - |r - u|}{r} u e^{-u^2/4\sigma^2} du$$

$$= 4\pi \left[\int_0^r \frac{u^2}{r} e^{-u^2/4\sigma^2} du + \int_r^\infty u e^{-u^2/4\sigma^2} du \right].$$

Define the error function by $\mathrm{Erf}\,z = (1/\sqrt{\pi}) \int_0^z e^{-x^2} dx$, so that $\mathrm{Erf}(\infty) = \frac{1}{2}$. Then a little calculation shows that N reduces to

$$N = 16\pi^{3/2}\sigma^3 \mathrm{Erf}(r/2\sigma)/r$$

and so $\phi = (2/r)\mathrm{Erf}(r/2\sigma)$. For small r this gives $\phi = (2/r)(r/2\sigma\sqrt{\pi}) = 1/\sigma\sqrt{\pi}$, that is, a high peak instead of the infinite value.

4. The difficulty in assessing Eddington's result lies in his basis of seeing the Coulomb potential as 'modified' by the uncertainty of the origin, whereas the Yukawa potential which he rejects is an addition to the Coulomb field. So that the question is of the difference between (going over to $z = r/2\sigma$) $1/z - 2\mathrm{Erf}(z)/z$ and e^{-pz}/z (where $p = 2\sigma k$ for the usual Yukawa $k = 1/r_0$). The problem is to determine the best value of p. Now firstly, consider small values of z, when $\mathrm{Erf}(z) = z/\sqrt{\pi}$, so that we are comparing $1/z - 2/\sqrt{\pi}$ and $1/z - p$. For small values of z, then, the value of p is $2/\sqrt{\pi} = 1.284$. An easy way to see what values should be taken for larger values of z is by a tabulation as shown in Table 12.1. It is clear that even taking $p = 1$ gives a good approximation for the inner range (as does $p = 1.128$) but there is a discrepancy in the range near to $z = 1$. To adjust this, note that for changes in $p\,\delta(e^{-pz}/z) = -e^{-pz}\delta p$ and so at $z = 1$ the required change of $0.0504 = \delta p/e$, giving $\delta p = 0.137$. The resulting value of 1.137 is close to the earlier value of 1.128. So taking $p = 1.128$ gives at once $r_0 = 1.773\sigma$.

Table 12.1

z	$[1 - 2\mathrm{Erf}(z)]/z$	e^{-z}/z
0		
0.001	999.2	999
0.01	99.2	99.005
0.1	9.204	9.048
1	0.3174	0.3678
10	0	0

References

Anderson C.D. 1929. *Zeit. für Phys.* **56**, 851.

Anderson C.D. and Neddermeyer S.H. 1936. *Phys. Rev.* **50**, 263.

Bastin E.W. and Kilmister C.W. 1954. *Proc. Camb. Phil. Soc.* **50**, 278.

Bastin T. 1966. *Studia Philosophica Gandensia* **4**, 77.

Bastin T., Noyes H.P., Amson J. and Kilmister C.W. 1979. *Int. Journ. Theor. Phys.* **18**, 445.

Benjamin W. 1968. *Illuminations* (trans. Harry Zohn, edit. Hannah Arendt). Harcourt Brace Jovanovich, New York.

Bond W.N. 1935. *Nature* **135**, 825.

Bondi H. 1960. *Cosmology*. Cambridge University Press, Cambridge.

Braithwaite R.B. 1940. *Mind* **49**, 455.

Bridgman P.W. 1922. *Dimensional Analysis*. Yale University Press, Cambridge, MA.

Cartan E. 1913. *Bull. Soc. Math. de France* **41**, 53.

Chandrasekhar S. 1931. *Mon. Not. Roy. Astr. Soc.* **95**, 207.

Chandrasekhar S. 1983. *Eddington, the most distinguished astrophysicist of his time*. Cambridge University Press, Cambridge.

Chinea F.J. 1989. *Gen. Rel. Grav.* **21**, 21.

Coleman A.J. 1945. *Phil. Mag.* (7) **36**, 269. (This is a summary of the main points in his Toronto PhD thesis of 1943.)

Dampier C.D. 1969. *The algebraic structure of Fundamental Theory*. London PhD thesis.

Dingle H. 1945. *Proc. Phys. Soc.* **57**, 244.

Dirac P.A.M. 1928a. *Proc. Roy. Soc.* (A) **117**, 610.

Dirac P.A.M. 1928b. *Proc. Roy. Soc.* (A) **118**, 351.

Dirac P.A.M. 1930. *Quantum Mechanics.* Clarendon Press, Oxford.

Dirac P.A.M. 1937. *Nature* **139**, 323.

Dirac P.A.M. 1938. *Proc. Roy. Soc.* (A) **165**, 199.

Douglas A.V. 1956. *Arthur Stanley Eddington.* Nelson, London.

Dyson F.J. 1952. *Phys. Rev.* **85**, 631.

Eddington Sir A.S. 1916. *Mon. Not. Roy. Astr. Soc.* **77**, 16; **77**, 596 and **78**, 113, (1917) and **79**, 22, (1918).

Eddington Sir A.S. 1917. *Observatory* **40**, 290.

Eddington Sir A.S. 1918. *Report on the Relativity Theory of Gravitation.* Physical Society, London.

Eddington Sir A.S. 1920a. *Proc. Phys. Soc.* **32**, 245.

Eddington Sir A.S. 1920b. *Mind* **29**, 145.

Eddington Sir A.S. 1920c. *Mind* **29**, 415.

Eddington Sir A.S. 1923. *Mathematical Theory of Relativity* (MTR). Cambridge University Press, Cambridge.

Eddington Sir A.S. 1926. *The Internal Constitution of the Stars.* Cambridge University Press, Cambridge.

Eddington Sir A.S. 1928a. *The Nature of the Physical World.* Cambridge University Press, Cambridge.

Eddington Sir A.S. 1928b. *Proc. Roy. Soc.* (A) **121**, 524.

Eddington Sir A.S. 1929. *Proc. Roy. Soc.* (A) **122**, 358.

Eddington Sir A.S. 1931a. *Jour. London Math. Soc.* **7**, 58.

Eddington Sir A.S. 1931b. *Proc. Roy. Soc.* (A) **133**, 315.

Eddington Sir A.S. 1933a. *The Expanding Universe.* Cambridge University Press, Cambridge.

Eddington Sir A.S. 1933b. *Jour. London Math. Soc.* **8**, 142.

Eddington Sir A.S. 1935. *New Pathways in Science.* Cambridge University Press, Cambridge.

Eddington Sir A.S. 1936. *Relativity Theory of Protons and Electrons* (RTPE). Cambridge University Press, Cambridge.

Eddington Sir A.S. 1939. *Philosophy of Physical Science.* Cambridge University Press, Cambridge.

Eddington Sir A.S. 1942. *The Combination of Relativity Theory and Quantum Theory,* Comm. Dublin Inst. of Advanced Studies, No. 2.

Eddington Sir A.S. 1944. *Proc. Camb. Phil. Soc.* **40**, 37.

Eddington Sir A.S. 1946. *Fundamental Theory.* Cambridge University Press, Cambridge.

Einstein A. 1905. *Ann. der Phys.* **17**, 891. (An English translation is to be found in *Special Theory of Relativity*, edit. C.W. Kilmister, Pergamon, Oxford, 1970.)

Einstein, A. 1916. *Ann. der Phys.* **49**, 769. (An English translation of part of this is to be found in *General Theory of Relativity*, edit. C.W. Kilmister, Pergamon, Oxford, 1973.)

Einstein A., Infeld L. and Hoffmann B. 1938. *Ann. Math.* **39**, 65.

Emden R. 1907. *Gaskugeln.* Teubner, Leipzig and Berlin.

Forman P. 1971. *Hist. Studies in the Phys. Sciences* **3**, 1.

Fowler R.H. 1926. *Mon. Not. Roy. Astr. Soc.* **87**, 114.

Good I.J. 1970. *Phys. Lett.* **33A**, 383.

Grattan-Guinness I. 1977. *Dear Russell – Dear Jourdain.* Duckworth, London.

Grieder K.R. 1984. *Found. of Phys.* **14**, 467.

Hamilton Sir W.R. 1844. *Proc. Roy. Ir. Acad.* **2**, 424. (This is the first publication but a transcript of Hamilton's notebook for the previous year is in Halberstam H. and Ingram R.E. 1967. *The Mathematical Papers of Sir William Rowan Hamilton.* Cambridge University Press, Cambridge.)

Hermann A., Meyenn K.v. and Weisskopf V. 1979. *Wolfgang Pauli: Wissenschaftlicher Briefwechsel*, Bd. 1 1919–29. Springer, Berlin.

Hiley B. and Peat F. 1987. (edit.) *Quantum Implications.* Routledge & Kegan Paul, London.

Jehle H. 1946. *Amer. Jour. Phys.* **14**, 60.

Kant I. 1787. *The Critique of Pure Reason* (there are many English translations).

Kilmister C.W. 1951. *Proc. Roy. Soc.* (A) **207**, 402. (This reproduces part of a London PhD thesis of 1950).

Kilmister C.W. and Tupper B.O.J. 1962. *Eddington's Statistical Theory.* Clarendon Press, Oxford.

Kilmister, C.W. 1992. *Philosophica* **50**, 55.

Klein F. 1872. *Vergleichende Betrachtungen über neuere geometrische Forschungen.* A. Düchert, Erlangen.

Lane J.H. 1870. *Amer. Jour. of Sci. and Arts* (2) **4**, 57.

Larmor, Sir J. 1900. *Aether and Matter.* Cambridge University Press, Cambridge.

Lattes C.M.G., Muirhead H., Occhialini G.P.S. and Powell C.F. 1947. *Nature* **159**, 694.

Lenz F. 1951. *Phys. Rev.* **82**, 554.

Levi-Civita T. 1926. *Math. Ann.* **97**, 291.

References

McGoveran D. and Noyes H.P. 1991. *Physics Essays* **4**, 115.

Merleau-Ponty J. 1965. *Philosophie et Theorie Physique chez Eddington.* Besancon, Paris.

Milne E.A. 1947. *Nature* **159**, 486.

Noyes H.P. and McGoveran D. 1989. *Physics Essays* **2**, 76.

Planck M. 1899. *Sitzber. Preuss. Akad. Wiss.*, p. 440.

Pirani F.A.E. 1956. *Acta Physica Polonica* **15**, 389.

Robb A.H. 1914. *The Absolute Relations of Time and Space.* Cambridge University Press, Cambridge.

Russell B. 1967. *The Autobiography of Bertrand Russell Vol. 1 1872–1914.* George Allen & Unwin, London.

Schrödinger E. 1950. *Space-time Structure.* Cambridge University Press, Cambridge.

Slater N.B. 1947. *Phil. Mag.* (7) **38**, 299.

Slater N.B. 1957. *The Development and Meaning of Eddington's Fundamental Theory.* Cambridge University Press, Cambridge.

Sommerfeld, A. 1916. *Phys. Zeit.* **17**, 491.

Stoner E.C. 1929. *Phil. Mag.* **7**, 63.

Stoner E.C. 1930. *Phil. Mag.* **9**, 994.

Tait, P.G. 1873. *Quaternions.* Clarendon Press, Oxford. (Second enlarged edition. The first edition, which is rather rare, was in 1867.)

Taylor J.G. 1987. *Tributes to Paul Dirac.* Adam Hilger, Bristol.

Temple G. 1930. *Proc. Roy. Soc.* **127**, 339; **127**, 349; **128**, 487 but most strikingly *Proc. Camb. Phil. Soc.* **26**, 496.

Tupper B.O.J. 1957. PhD thesis, London.

Wigner E.P. 1957. *Rev. Mod. Phys.* **29**, 255.

Wyler A. 1969. *Compt. Rend. Acad. Sci.* (Paris) **269A**, 743.

Wyler A. 1971. *Compt. Rend. Acad. Sci.* (Paris) **271A**, 186.

Yukawa H. 1935. *Proc. Phys. Math. Soc.* (Japan) **17**, 48.

Index

acceleration field, 28
algebra, 101, 105
Anderson, 103
astrophysics, 21–23, 103, 191
atomicity, 201

Bastin, xi, 52
Benjamin, 245
black-body radiation, 80
Bohr, 104, 113, 116
Bond factor, 191, 199, 234
Bondi, 56, 104
Born, x, 84
Born interpretation, 135, 160
Braithwaite, 57, 244
Bridgman, 51, 243

Cartan, 76
Cepheids, 21
Chandrasekhar, 103, 191
Clifford algebra, 192, 216–17, 227
Coleman, 162, 164
combined system, 160, 167, 169, 211
commutation rules, 86
compact E-number 140
comparison, 127, 138
comparison, fluid, 139, 142, 145–6, 149
Compton wavelength, 114, 197
conditions of the world, 51, 118
conservation of probability, 134
continuity equation, 45
correspondence principle, 83
Coulomb potential, 125, 132, 149
Crowther, 83

curvature, 30, 165, 192, 194–5, 208

Dampier, 6, 227, 232–3
Darwin, 94, 105, 112, 139, 224
de Broglie, 86
de Sitter, 27
de Valera, 223
degeneracy, 114
degrees of freedom, 116
descriptive tolerance, 20, 49, 208
differential geometry, 29
Dingle, 43, 51
Dirac, x, 2, 4, 5, 7, 55–6, 65, 79, 85, 90, 93, 101, 104, 117, 124, 130, 225, 231
direct product, 170
double frame, 112
dynamical variable, 92
Dyson, 13, 35, 115

eclipse expedition, 34–5
EF-algebra, 112, 149
Einstein, x, 14–15, 17, 26, 28, 36, 45–6, 66, 82, 159, 188
Einstein universe, 54, 193, 195
electric charge, 124
electromagnetism, 47
Emden, 22
energy tensor, 60
E-numbers, 110
environment, 139, 172, 236
epistemological positivism, 117
epistemology, 216
exclusion, 126, 128
external particle, 136

falsifiability, 49, 207
Fermi-Dirac statistics, 127, 129, 149
field equations, 33
fine-structure constant, xi, 4, 113, 124, 146, 189–90, 198, 214, 231
Forman, 117
fundamental theories, 235

Gauss, 29, 141
general relativity, 26
geodesics, 23
Good, 214
Gordon, 92
gravitational red-shift, 36
Greenwich observatory, 13
Grossmann, 29
group representation, 70

Hallwachs, 82
Hamilton, 108
Heisenberg, x, 84
Heitler, 225
Hertzsprung and Russell, 22
Hicks, 2
Hiley, 111
Hoffman, 45
Hubble constant, 197
hydrogen atom, 136

indistinguishability, 126–7, 149
indistinguishable particles, 132
inertial frame, 27
Infeld, 45
interchange, 127, 130–1, 146, 149
interchange momentum, 137
internal particle, 136

Jeans, 22, 83
Jordan, x, 84

Kant, 37, 45, 49–50, 58, 109, 231
Kepler problem, 33, 149
Klein, 66, 116

Lamb, 11, 190
Lane, 22
Laplace, 44
large numbers, 55, 118, 157
Larmor, 16
laws of nature, 52
Lemaître, 201
Lenz, 213
Lorentz, 15, 135
Lorentz transformation, 111

Mach's Principle, 159

Mann, ix
mass-ratio (of proton/electron), 189–190, 198, 236
material reference frame, 138
matter, 42
Maxwell, x, 5
measure, 201
Merleau-Ponty, 6, 227, 229–30
Mercury, 34
metric, 30
Michelson, 16–17
Milne, 3, 225

Nernst, 22
new quantum theory, 84
non-Coulombian potential, 236–9
non-redundancy, 61, 115, 208
number of particles, 190, 200

observable, 237
observer participation, 211
old quantum theory, 4, 80
operationalism, 49, 208
Osborne, 211

parallel displacement, 31, 141
Pauli, 104, 116, 128, 189
Peat, 111
perception, 43
perfect object system, 237
phase, 161
phase-space, 160, 162, 164, 174, 210, 228
photo-electric effect, 82
Planck, x, 81, 243
Poincaré, 15
Popper, 49
principle of equivalence, 28
principle of identification, 59, 208
principle of relativity, 65
probability, 85, 175
proton mass, 156

q-number, 85
quadratic equation (for mass-ratio), 137, 158, 177, 199
quantum mechanics, 79
quantum theory, 58
quaternions, 108

Rayleigh, 81
recoil, 127, 142, 144
relativistic degeneracy formula, 103, 191
Riemann–Christoffel tensor, 31, 140, 143, 149
Robb, 59
Römer, 19

Index

Rosenfeld, 104
Russell, 57, 188
Rutherford, 18
Rydberg constant, 229

scale, 236
scale albebra, 238
scale constants, 212–3
Schrödinger, x, 76, 86, 135, 223
Schwarzschild, 34
selective subjectivism, 20, 48, 207
sharp spectral lines, 80
Slater, 6, 226–7, 231
Sommerfeld, 114, 116
special relativity, 14–21
spectroscopy, 82
speed of light, 5
spinors, 72, 74
standard mass, 197
state, 84–5
stellar structure, 4
stereographic projection, 166
Stoner, 103
structure, 57, 118, 208
superposition, 130

surface geometry, 29

Tait, 108
Temple, 189
tensor identities, 192
tensor representations, 71
theory-language, 52, 55, 133, 208
three-dimensionaltiy (of space), 8, 109
Tupper, 6, 227, 236, 240–1
two-valued representations, 74

uncertainty of origin, 197, 211, 228, 238
uncertainty principle, 89
unified field theories, 47

vector representations, 71

Weyl, 47, 66–70
Whittaker, 5, 202, 224, 226
Wien, 81
Wittgenstien, 49
Wyler, 214

Yukawa, 178